ADVANCED LEARNER'S DICTIONARY *Of* CHEMISTRY

ADVANCED LEARNER'S DICTIONARY

Of

CHEMISTRY

Compiled & Edited by

A Team of Experts

ANMOL PUBLICATIONS PVT. LTD.

NEW DELHI - 110 002 (INDIA)

ANMOL PUBLICATIONS PVT. LTD.
4374/4B, Ansari Road, Daryaganj
New Delhi - 110 002

Advanced Learner's Dictionary of Chemistry

First Edition, 2000
Reprint : *2001*
ISBN 81-261-0462-7

PRINTED IN INDIA

Published by J.L. Kumar for Anmol Publications Pvt. Ltd., New Delhi and Printed at Mehra Offset Press, Delhi. *Phone* : **326 85 74**

Preface

Chemistry is the branch of physical science concerned with the composition, properties and reaction of substances. Inorganic Chemistry, Organic Chemistry, and Physical Chemistry are the major branches of chemistry.

Basically, this dictionary is meant for the students at advanced level, teachers, scientists, chemists, chemical engineers, and chemical analysts. All the entries are judiciously selected by a team of experts in the field in succinct manner. Different chemical formulae are explained with the help of suitable diagrams.

We are grateful to all learned scientists whose definitions and explanations are borrowed/substantially made use of in this work.

It is hoped that the students, teachers, and the general users will find this endeavour unique, informative and useful. For making the dictionary more comprehensive and informative, constructive criticism from the readers are welcome.

—Editors

A

Abrasive: A substance used for grinding or rubbing down surfaces *e.g.* emery.

Absolute Alcohol: Obtained by digesting rectified spirit over quick lime for several clays and then redistilling.

Absolute Expansion: The extent to which a liquid expands on heating.

Absolute Temperature: Temperature measured on a scale in (Kelvin scale) which the zero lies at minus 273 degrees celsius.

$$°C + 273.15 = K$$

Absolute Zero (–273.15°C): The temperature at which parti-cles would lose all their kinetic energy, but which can never be reached in practice.

Absorption: A process in which gas is taken up by a liquid or solid, or a liquid is taken up by a solid. It is a bulk on phenomenon, *i.e.*, the absorbed material is uniformly distributed throughout the body of the substance.

Absorption Indicator: An indicator used for librations that involve a precipitation reaction. The method depends upon the fact that at the equivalence point there is a change in the nature of the ions absorbed by the precipitate particles. Floures cain a florescent compound is commonly used. For example, in the libration of sodium chloride solution with added silver nitrate, silver chloride is precipitated, Sodium ions and chloride ions are absorbed in the precipitate. At the end point silver ions and nitrate ions are in sight excess and silver ions are then absorbed.

Absorption Spectroscopy: A spectroscopic technique, qua-litative or quantitative, depending upon the measurement of an absorption spectrum.

Abundance: The relative amount of a given element as compared with other elements, *e.g.*, oxygen is by far the most abundant element in the earth's crust—of the mass of the earth's crust 49.5% is due to oxygen atoms. Silicon, the next most abundant element is also as plentiful.

Acaricidal, carbinols: Acaricidal derivatives of 1, 1-di (4(chlorophenyl) metha-nol; 1,1-bis(4-chlorophenyal) ethanol (chlorofenthol); 1,1-bis(4-chlorophenyl) cyclopropylmethanol (prochol-nol); 1,1-bis (4-chlorophenyl)-2,2,2-trichloroethanol (dicofol), 1,1- bis (4-chlorophenyl)1-ethoxycarboxymethanol (chloro-benzilate) *etc.* They are powerful acaricides, but differ in their bioactivity spectra.

Accelerator: A substance which increases the rate of reaction. Also a device in automobiles used to increase the speed.

Acceptor: A group or atom to which a pair of electrons is donated in a coordinate bond.

Accumulator: Lead storage battery used to store electricity. It gets charged when an electric current passes through it.

Acetal: An organic compound obtained when alcohol is added to the carboxyl group of aldehyde in the presence of anhydrous acids.

Acetanilide or N-Phenyl-acetamide ($C_6H_5NHCOCH_3$): A colourless crystalline (rhombic plates) substance (m.p. 387 K), sparingly soluble in cold water but more soluble in hot water. It is obtained by the action of glacial acetic acid on aniline.

$$C_6H_5NH_2 + CH_3COOH \rightarrow C_6H_5NHCOCH_3 + H_2O$$

Acetate: Acetates formed from the and elthouoic acid. It is a fragrant liquid used as a solvent for plastic and in flawing and perfumery.

Acetic Anhydride: A colourless liquid, b.p. 139.50°C, having irritating smell. It is obtained by distilling a mixture of anhydrous sodium acetate and acetyl chloride.

$$CH_3COONa + CH_3COCl \rightarrow (CH_3CO)_2O + NaCl$$

Acenaphthene (C12H10): A colourless needle-shaped solid having m.p. 95°C. It is used as a dye stuff interme-diate.

Acetonitrile or Methyl Cyanide (CH_3CN): A colourless poisonous liquid, b.p. 82°C, with an odour like ether. It is used as organic solvent.

Acetophenone or Methyl Phenyl Ketone ($C_6H_5COCH_3$): A colourless crystalline solid (m,p. 293K) obtained by treating benzene with acetyl chloride, in the presence of an hydrous aluminium chloride. It has a pleasant smell, is slightly soluble in water, is used to prepare hypnone (sleep producing drug) and in perfumery.

Acetylene: A gaseous alkyne, C_2H_2. Used in oxyacetylene welding torch.

Acetylene black (Caprene): A type of carbon black obtained by pyrolysis of ethyne.

Achiral: The molecule of a compound which is not optically active.

Achromatic Indicators: Substances which give a grey endpoint and generally find application in the titration of turbid liquids.

Acids: Chemicals that produce positively charged hydrogen ions (H^+) or a proton in solutions. They are neutralized by bases and form salts. Strong acids such as hydrochloric acids readily dissociate in solutions giving out hydrogen ions.

Acid Anhydride: A type of organic compound of general formula RECOOCOR,

where R′ and R are alkyl or aryl groups. They are prepared by reaction of an acyl halide with the sodium salt of a carboxylic acid, *e.g.*,

$$RCOCl + R'COONa\ RCOOCOR' + NaCl$$

Acid (Lewis): Compound, capable of sharing an electron pair form a Lewis base, *e.g.*, aluminium chlo-ride, boron trifluoride, ferric chloride, zinc chloride etc. More recently any chemical compound capable of accep-ting an electron in an acid and that which can donate an election is a base.

Acidic: Having a tendency to release a proton or to accept an electron pair from a donor. In aqueous solution the pH is a measure of the acidity, *i.e.*, and acidic solution is one in which the concentration of H_3O^+ exceeds that in pure water at the same temperature,. *i.e.*, pH is lower than.

Acid Base Indicator: A substance, weak acid or weak base which has different colour in acid or base solution. The colour change is due to a marked difference in colour between the undissociated and ionic. For a good indicator, the colour change must occur between limits of p^H, *e.g,* methyl orange is red at p^H3.1 and changes to yellow at p^H 4.4.

Acid, carboxylic: Organic compound containing a carboxylic function (–COOH group) e.g., acetic acid, benzoic acid etc.

Acid Dyes: A group of dyes, nearly all salts of organic acids used for dyeing natural silk and wool.

Acid Halide: An organic compound in which the OH group of a carboxylic acid is removed by a halogen. A simple acid halide derived from acetic acid is acetyl chloride,

$$CH_3-\overset{\overset{\displaystyle O}{||}}{C}-Cl$$

Acid Rain: Rain in which the water is little acidic than usual. This happens because of dissolving of certain air-Pollution like. Sulphur dioxide, nitrogen oxides and carbon monoxide in the rain water. Acid rains are concern in heavily polluted industrial areas and often damage the environment by affecting the plants and fertility of the soil.

Acidic hydrogen: A hydrogen atom in a molecule that enters into a dissociation equilibrium when the molecule is dissolved in a solvent. For example, in acetic acid (CH_3COOH) the acidic hydrogen is the one on the carboxyl group—COOH.

Acid Salt: The compound obtained after replacing one hydrogen atom by a metal ion, *e.g.,* $NaHCO_3$, $NaHSO_4$ *etc.*

Acid Sludge: A complex acid residue produced when kerosene lubricating oil or other petroleum products are refined with sulphonic acid or oleum. It consists of hydrocarbons, sulphuric acids and free sulphuric acid. Petroleum sulphonates can be recovered by extraction with alkali and used for the metal processing, textile and leather industries.

Acid Value: The number to neutralise 1 gram of oil or fats, indicates the amount of free acid present.

Acid Dissociation Constant: Most of the substances that are acidic in ionizes in aqueous medium can be expressed by the equilibrium constant for the ionization reaction. In general we can represent any acid by symbol HX, where X^- is the formula for the conjugate base that remains when the proton ionizes.

Aconitic Acid ($C_6H_6O_6$): An unsaturated acid obtained by dehydrating citric acid with 50% sulphuric acid.

Acrolein (CH2=CH.CHO): Acrylic aldehyde or acryla-dehyde, a molluscicide against rivers snails. It is sufficiently water soluble but lachrymatory.

Actinium: The silvery metal belonging to group 3 glows in the dark owing to its radioactivity. It was discovered in the residue of pitchblende after radium extraction. It is extracted from uranium or lanthanu more or artificially produced by bombarding radium with neutrons.

Symbol : Ac; a.n. 89; m.p. 105°C; b.p. 3200°C; r.d. 10.07.

Activated Charcoal: A powdery form of charcoal. It is highly porous and can absorb a good amount of gases. It is obtained by driving away air filling the pores of common charcoal by heating. It is used in industries for filtering out gases.

Activated Complex: The activated complex is the particular arrangement of reactant and product molecules at the point of maximum energy in the rate determining step of a reaction.

Activation Energy (Symbol Ea): The minimum energy that must be given by the reactants in a chemical reaction to overcome the barrier to formation of products.

Active Series: In the active series metals are arranged in order of their ease of their oxidation. The metal on the top are most easily oxidised. The active series indicates not only the reactivities of metals with acids but also their reactivities with each other.

For example, the position of copper in the activity series indicates that it has a strong attraction for electrons than does zinc because it is more easily reduced. Conseque-ntly, copper would end up with electrons. In any oxidation-reduction between a metal and a metal ions the more active will end up in the oxidized state.

Actionic Radiation: Radiation that can cause a thermical reaction, for example, ultraviolet radiation is actionic.

Activity: The activity of a substance may be thought of as an effective concentration. It has the property that in all situations the equilibrium constant expression can be written as

$$\frac{[C]^c\,[D]^d}{[A]^a\,[B]^b}$$

where the notation [A] stands for the activity of A. Thus, in terms of the activity, the equilibrium-constant expression has its simple form and is always constant at a particular temperature. The activity of an ideal gas is equal to the pressure in atmospheres but for nonideal gas the activity must be calculated.

Acrilon: A trade name for a synthetic fibre, based on a copolymer of acronitrile with small proportion of other unspecified vinyl monomers.

Acrylic Acid [CH_2:CH. COOH): A corrosive liquid b.p. 141°C. Derivatives from the basis of the important acrylic resins.

Acyclic or Open Chain: Organic compounds which are not cyclic, *e.g.*, saturated hydrocarbons (methane, ethane propane) and unsaturated hydrocarbons (ethylene acetylene).

Acylation: An addition of RCO group to the aromatic ring.

Acyl Group: The group of atoms RCO—

Acyl Halide: A kind of organic compound with general formula,

$$\begin{array}{c} O \\ \| \\ R\text{–}C\text{–}X \end{array}$$

where X stands for chloride, bromide and iodide. It may be prepared by the replacement of the hydroxyl in the carboxyl group by chlorine, bromine and iodine.

Activity Coefficient: Symbol of arcause of the degree of decoration from ideality of a solid substance as $a = fc$.

where a is the activity and c the concentration. For an ideal solute $f = 1$; for real system can be less or greater than unity.

Add Reaction: A reaction in which two molecules of aldehyde combine to give an aldd -e, e, a compound containing both aldehyde and alcohol functional group. The reaction is base-catalysed; the reaction of ethanol with sodium hydro-xide gives :

$$2CH_2CHO = CH_3CH(OH)CH_2\,CHO$$

The mechanism is singular to that of the class conden-sation; the first step is removal of proton to give a carbonion, which subseque-ntly attacks the of the carbonyl group on the other molecule

$$CH_3CHO + OH^- = \text{-}CH_2CHO + H_2O$$

Addition Reaction: A reaction in which two molecules combine to yield a single molecule product.

Addition Polymerization: Addition polymerization occurs where the moments undergo addition reactions, with no other substance formed.

Adiabatic Changes: A change deriving which no energy enters or leaves the system.

In an adiabatic expansion of a gas, mechanical work is done by the gas as its volume in veaves and the gas temperature falls. For an ideal gas rider going a reversible adiabatic change it can be shown that

$$pV^{\gamma} = K_1$$

$$T^{\gamma}p^{1-\gamma} = K_2$$

$$\text{and } TV^{\gamma-1} = K_3$$

where K_1, K_2 and K_3 are constants and γ is the ratio of the principle specific heat capacities.

Adduct: An addition compound obtained by a reaction in-volving no valency changes.

Adipic Acid ($CO_2H.(CH_2)_4\ CO_2H$): A crystalline solid m.p. 150°C. It is prepared in the industries by the oxidation of cyclohexanol with concentrated nitric acid in the presence of ammonium vandate as catalyst. It is used for the preparation of resins (polyesters) and is an intermediate in the manufacture of nylon.

Adhesive Forces: The forces between unlike substances, such as water rid glass. In a glass tube, the adhesive forces between water and glass re sufficiently stronger than the adhesive forces of water to form a concave—upward surface.

Adonic Acid: An acid drived from an aldose by oxidation of the aldehyde roup to a carboxyl group, *e.g.*, glu-conic acid.

Adsorption: The particles at the surface of a solid or liquid are in a state of tension and in order to satisfy their residual valencies exhibit a tendency to attract on to and retain at their surface other particles with which they come into contact.

Aerobic fermentation: Biochemical reaction by microbes and bacteria in presence of oxygen *e.g.*, fermentation of sugar to alcohol and alcohol to acetic acid.

Aerogel: A highly porous lightweight solid obtained by drying out all the liquid in a gel.

Aero Metal: A casting alloy consisting mainly of alumi-nium, zinc and copper.

Aerosol: A dispersion of solid or liquid particles in a gas, *e.g.*, (solid-in-gas aerosol) and mist (liquid-in-gas aerosal).

Agar: A gelatine-like material obtained from sea weeds. It is an important constituent of tissue culture medium on which bacteria are grown.

Agate: A very hard natural form of silica. It is used for knife-edges of balances, for mortars and in ornaments.

Agglutiration: The proves in which particles suspended in a liquid clamp together, *e.g.*, clumping together of blood call or bacteria.

Aireation: Oxygen is an effective purifying agent. It attacks many forms of bacteria and hastens oxidation of dissolved organic, materials. To increase the concentration of dissolved oxygen, the water may be sprayed into air.

Air Hardening: A method of hardening steel by allowing it to cool naturally in air or an air blast.

Alcohols: A type of organic compound of the general formula ROH, where R is a hydrocarbon group. Example of simple alcohols are momentral (CH_3OH) and ethanol (C_2H_5OH)

Alabaster: Natural opaque form of hydrated calcium sulphate $CaSO_4{,}2H_2O$.

Alchemy: Modem chemistry emerged from alchemy. There was much more to alchemy than the mere quest for the ability to transform base metals into gold. Alchemy was a complex mixture of philosophy, astrology, mysticism, magic, real chemistry and other ingredients including the belief that all forms of matter have a soul in common. From the modem point of view, alchemy was valuable because it led to the development of laboratory procedures and to the discovery of new substances.

Alcoholmetry: The process of determining the amount of ethanol in a liquid.

Alcoholysis: A reaction in which an alcohol plays a part similar to that of water in hydrolysis.

Aldehydes: A kind of organic compound with general for-mula.RCHO, where the -CHO group consists of a carbonyl group attached to a hydrogen atom.Examples are methanal and ethanal.

Aldrin: The insecticidal pro-duct containing not less that 95% by weight of 1, 2, 3, 4, 10, 10, hexachloro 1, 4, 4a, 5, 8, 8a, hexahydro—exo 1, 4, -endo, 5, 8-dimethanona-phthalene. It is prepared from Diels-Alder addition product of cyclopentadiene and vinyl chloride by dehydrochlorination follo-wed by condensation with hexachlorocyclopentadiene. It is toxic to humans and animals.

Aldol (acetadol, b-hydroxy-butyraldehyde, 3-hydroxy-butanal): A colourless liquid, b.p. 83°/20 m.m., miscible with water and ethanol. It is prepared from acetaldehyde in the presence of NaOH.

Aldoximes: The compounds formed by the action of hydroxylamine on aldehydes containing the group—CH= NOH

Aldohexose: An aldose sugar with six carbon atoms.

Aldopentose: An aldose sugar with free carbon atoms.

Aldose: A sugar containing an aldehyde (CHO) or potential aldehyde group.

Alginic Acid [(C6H8O6)n]: A complex organic compound linked to the carbohydrates. It occurs in some seaweeds. It is used for preparing emulsions and as a thickening agent in food industry.

Alicyclic: Aliphatic cyclic compounds, *e.g.*, cyclohexane.

Aliphatic Compounds: An organic compound with proper-ties similar to those of the alkanes, alkenes, alkynes and their derivatives.

Alizarin (1: 2 -Dihydroxy anthraquinone): A ruby red crystal (m.p. 560 K) obtained by sulphonating anthraquinone with oleum at 430 K and fusing with sodium salt of the product with sodium hydroxide and calculated quantity of potassium chlorate at 450470 K under pressure. It is used in dyeing with the help of mordants.

Alkyl Group: A group formed by removing hydrogen atom from an alkane or other aliphatic derivatives.

Alkali: A water soluble strong base. They are very corrosive and react with acids, *e.g.*, NaOH.

Alkali Metals (1-group ele-ments): The alkali metals (Li, Na, K, Rb, Cs, Fr) are never found in the elemental state in nature, for they react rapidly and completely with virtually all nonmetals. They are very soft metals, having bright silvery lustre and are good conductors of electricity and beat. The great chemical reactivity of the alkali metals is a reflection of their relatively low first ionization energies.

Alkaline-Earth Metals: The alkaline metals (Be, Mg, Ca, Sr, Ba, Ra) are never found in the metallic form in nature, because like alkali metals, they are active reductants and react readily with a variety of nonmetals. They can be prepared by electrolysis of their fused halides. These metals are all considerably harder than alkali metals.

Alkalimetry: Volumetric analysis or acid base titration in which a standard solution of an acid is added to the unknown (base) solution plus the indicator. Alkalimetry is the converse, i.e., the base is in the burette.

Alkaloids: A group of water insoluble crystalline chemicals produced by plants and animals. The opium alkaloids, for example, are groups of more than 20 amines found in resin. The most abundant of the opium alkaloids is morphine. Morphine and many of its numerous derivatives such as codeine, are extremely effective pain killers.

Alkanes: Many of these substances are familiar because of their widespread use. Methane is a major compound of natural gas and is used for home heating in a gas stove and hot water heater. Propane is the major component of bottled, or LP, gas used for home heating, cooking. The general formula of alkane is $C_nH_{2n}+2$ and their type formula is RH. The molecular formula of the various members of the homologous series can be obtained by putting n = 1, 2, 3, 4, *etc.*

Alkene (Olefins): The unsaturated hydrocarbons that contains one double bond with the general formula C_nH_{2n}. The members containing two, to four carbon atoms are gases; five to fifteen liquids, sixteen onwards solids. All are lighter than water, in which they are insoluble and they burn in air with a luminous smoky flame. They are obtained in huge quantity in the cracking of petroleum and are used to prepare large number of organic compounds.

Alkoxide: An organic compound formed when alcohols are attacked by strongly electropositive metals, *e.g.*, sodium and potassium, hydrogen is liberated, *e.g.*, ethanol acts with sodium to form ethoxide.

$$2C_2H_5OH + Na \rightarrow \underset{\text{sodium ethoxide}}{2C_2H_5ONa} + H_2$$

Alkyl Benzene: A type of organic hydrocarbon containing one or more alkyl groups substituted onto a benzene ring. Methyl ben-zene ($C_6H_5CH_3$) and 1, 3-dimethyl benzene are simple examples. Alkyl benzene can be made by the Friedel-Gafts reaction or by the Fitting reaction. Industrially, the hydroforming of crude oil.

Substitution of alkyl benzenes can occur at the benzene ring. The alkyl group directs the substituent into the 21 or 4 position.

Substitution of hydrogen into the 2 or 4 position. Substitution of hydrogen atoms on the alkyl group can also occur.

Alkyl group (radical): Univalent hydrocarbon group of general formula C_nH_{2n}, e.g., methyal (CH_3—); ethyl (C_2H_5—) etc. formed generally by removal of a hydro-gen atom from an alkane molecule.

Alkyl Halides: The organic compounds with general formula RX, where R is any simple alkyl or substituted alkyl group and X may be fluoride, chloride, bromide and iodide. In of their polarity, alkyl halides are not soluble in water, probably because of their inability to form hydrogen bonds.

Alkynes: Organic compounds commonly called acetylenes, characterized by the presence of a triple bond between the two carbon atoms. The acetylenes formula C_nH_{2n-2} and triple bond is also known as "acetylenic bond".

Alloy: A metal prepared by adding other metals or nonmetals to obtain desired properties. Brass is an alloy of copper and zinc. Some kinds of steel are alloys of iron and carbon and can be considered as solid solutions in which carbon atoms are located in some of the spaces between iron atoms.

Allotropy: The ability of certain elements to exist more than one physical form Allotropy is very common in groups IV, V and VI of the periodic table than in other groups. Graphite and diamond are the two important allotropic forms of carbon. Sulphur also exists in various allotropic forms like rhombic, monoclinic, plastic and sulphur etc.

Allyl Alcohol (CH_2:$CHCH_2OH$): A colourless pungent liquid, b.p. 96.5°C. It is used in the manufacture of synthetic resins and pharmaceuticals.

Aloxan [$(CO)_4$ $(NH)_2$]: A white crystalline hetero-cyclic compound m.p. 170°C, derived from uric acid by treatment with dilute nitric acid.

Alpha Decay: A form of radioactive decay in which a nucleus spontaneously emits an alpha particle.

Alpha Particle: The nucleus of a helium atom consisting of two protons and two neutrons, it has a relative atomic mass of 4.2029 and a charge of +2.

Alum: A double sulphate of aluminium and alkali metals with 24 molecules of water, *e.g.*, potash alum K_2SO_4, $Al_2(SO_4)_3$.$24H_2O$ and soda alum Na_2SO_4.$A1_2(SO_4)_3$. $24H_2O$. Most of other trivalent metals like Fe, Mn, Ga, Co, Cr, V, *etc.*, are also found to form similar double sulphates.

with monovalent metals or radicals like Na, K, Rb, Cs and NH_4^+. Different alums are named differently according to the metals constituting them, *e.g.*, $Na_2SO_4 \cdot Cr_2(SO_4)_3\, 24H_2O$ is sodium chromium alum and $K_2SO_4 \cdot Fe_2(SO_4)_3 \cdot 24H_2O$ is potassium iron alum.

Alums are used in tanning of leather and as mordants of dyeing. They are also employed as coagulants and for antiseptic purpose in medicine.

Aluminium: A light silvery white metal. Good conductor of, electricity. It is recovered pure from its oxide by electrolytic reduction. Aluminium foil is used for packing fine articles, *e.g.*, sweets, photographic and pharmaceutical products. Aluminium cables are preferred to those of copper on account of their lightness and low cost. Aluminium and its alloys are light and durable. Therefore, they are used in automobile and aircraft industries.

Symbol : AI; a.n, 13; m.p. 659.8°C; b.p. 180°C, r.d. 2.702.

Aluminium Acetate ($AI(CH_3COO)_3$): A white salt obtained by mixing the solutions of aluminium sulphate and lead acetate.

$$Al_2(SO_4) + 2Pb(CH_3COO)_2 \rightarrow 2Al(CH_3COO)_3 + PbSO_4\downarrow.$$

The precipitate of lead sulphate is removed by filtration and the aluminium acetate remains dissolved in the filtrate. Any attempt to evaporate the solution results into the formation of basic acetate due to hydrolysis. Its solution is used in medicine as disinfectant and antiseptic. It is also used to prepare waterproof cloth and as a mordant.

Aluminium Bimethyl [$(CH_3)_3Al$]: A colourless liquid produced by the sodium reduction of dimethyl aluminium chloride.

It ignites spontaneously on contact with air and reacts violently with water, acids, halogens, alcohols, and amines. Aluminium alloys are used in the Ziegler process for the manufacture of high density poly effective.

Aluminium Bromide ($A1Br_3$): A white solid soluble in water and many organic solvents.

Aluminium Potassium Sulphate: (Potash above $Al_2\,(SO_4)_3K_2SO_4.24H2O$, $AlK(SO_4)_2.12H_2O$): A white solid, soluble in water but insoluble in alcohol, prepared by mixing dequim molecular quantities of solutions of one moment and aluminium sulphate followed by crystallization.

Aluminium Chloride ($AlCl_3$): A white hydroscopic solid that fumes in moist air. Its vapour density at 350-400 corresponds to a bimolecular formula Al_2C_6. On heating to 800°C, it converts back to $AlCl_3$

$$Al_2C_6 \xrightarrow{800^\circ} 2AlCl_3$$

It is highly soluble in water and the aqueous solutions are strongly acidic on account of the hydrolysis. It is widely used in dyestuff industries and for the cracking of petroleum. It is also used as catalyst in organic synthesis.

Aluminium Fluoride (AlF_3): A colourless substance obtained by the action of HF on $Al(OH)_3$. An inert material, relatively insoluble in most solvents unattacked by acid and alkali.

Aluminium Hydroxide [$Al(OH)_3$]: A white gelatinous solid obtained by adding ammonia to the solution of aluminium salts. It is used to prepare waterproof cloth and as a mordant.

Aluminium Nitrate [$Al(NO_3)_2$]: A white solid obtained by adding freshly precipitated aluminium hydroxide in dilute nitric acid, or by mixing solution of aluminium sulphate and lead nitrate and filtering off the lead sulphate. It is used as mordant for dyeing and preparing gas mantles.

Aluminium Nitride (AlN): A yellow crystalline substance obtained by heating aluminium in a current of dry nitrogen or ammonia. It may also be obtained by heating a mixture of bauxite and carbon to 1600°C in a current of dry nitrogen.

$$Al_2O_3 + 3C + N_2 \rightarrow 2AlN + 3CO$$

It is decomposed by water or hot alkalis forming hydroxide with evolution of ammonia.

Aluminium Silicates: A large class of minerals, both natural and synthetic, having aluminium and silicon combined with oxygen in their structure.

Aluminium Sulphate $Al_2(SO_4)_3$: Anhydrous aluminium sulphate is a white powder. It is fairly soluble in water, giving an acidic solution because of hydrolysis. It is used to manufacture alums, as a refill in foam type fire-extinguishers, tanning of leather, as mordant for dyeing and a mixture of sodium bicarbonate and aluminium sulphate is used as baking powder.

Aluminium Sulphide (Al_2S_3): Prepared by fusing alumi-nium powder with sulphur.

$$2Al + 3S \rightarrow Al_2S_3$$

It reacts with water to form aluminium hydroxide and hydrogen suphide.

$$Al_2S_3 + 6H_2O \rightarrow 2Al(OH)_3 + 3H_2S$$

Alundum: An artificial form of Corundum (a-Al_2O_3) made by fusing calcined bauxite in an electric furnace. It is used for highly refractory bricks, crucibles and muffles.

Alunite or Alum Stone A natural compound of potassium and aluminium sulphate and aluminium hydroxide [$K_2SO_4.Al_2SO_4$. $4Al(OH)_3$]. It is a source of alum.

Amalgam: An alloy of a metal with mercury.

Amalgous: It may be liquid or solid. An and gone of sodium (Na/Hg) with water is hydrogen.

Amatol: An explosive mixture of 80% ammonium nitrate and 20% T.N.T.

Americium: A silvery white element belonging to group 3, obtained by neutron irradiation of pure Pu ^{241}Am is a valuable laboratory source of a-particle. It is used in radiography as a source of gamma rays. Symbol : Am; a.n. 95; b.p. 2607°C m.p. 994°C r.d. 13.67.

Amide: A kind of organic compound of general formula $RCONH_2$. Amides are white solids, which can be formed by the partial dehydration of the ammonium salt of a carboxilic acid.

Amination: The conversion of aldehyde or ketone into amine by reacting them with hydrogen and ammonia in the presence of a catalyst.

Amines: Organic compounds with the general formula RNH_2, R_2NH, or R_3N, where R is any alkyl or aryl group.

Amino acid: A carboxylic acid which contains an amino group (-NH_2) in the alkyl side chain. These are classified into (a, b, g, or d amino acids depending on whether the amino-group is on 2nd, 3rd, 4th or 5th carbon from the carboxylic groups. These are the basic units which combine to form peptide,— $CH(NH_2)CONH\ CH_2$—or polypeptide, —NHCHR CONHNCH(R) CONHCH(R)CONH— chain in proteins. There are 24 alpha aminoacids that occur in nature.

Amino Acids: Twenty amino acids are found to occur commonly.

Organic acids which are constituents of proteins in all organisms. All of these are a-aminoacids; that is, the amino-group is located on the carbon atom immediately adjacent to the carboxylic acid functional group. 2-

Aminophenol (C_6H_7NO): A colourless crystalline solid having M.P. 174°C. It is obtained by reduction of 2-nitrophenol with sodium sulphide and used as a fur and hair dye.

Aminoplastic Resins: The synthetic resins obtained from the reaction of urea, melamine, or allied amino compounds with aldehydes. They form the basis of thermosetting moulding material.

Ammine: A complex in which ammonia Molecules are coordinated to a metal ion *e.g.*, $[Cu(NH_3)_4]^{2+}$.

Ammonal: A mixture of ammonium nitrate and aluminium. It is used as an explosive.

Ammonia (NH_3): The gas has a sharp, pungent odour which makes it possible to detect even very smal: concentrations. Although it is not a poisonous gas, ammonia has a strongly irritating action on the mucous membranes of the eyes, nose, throat and lungs and the gas in high concentration may cause severe and permanent injury to the eyes. It is obtained commercially from the combination of hydrogen and nitrogen.

The chief applications are the manufacture of other nitrogen compounds particularly nitric acid, and ammonium salts. Aqueous solution of ammonia because of its mild alkalinity, is a familiar household cleaning agent. Liquid ammonia serves as a good refrigerant in iceplants.

Ammonium Acetate (CH_3COON_4): A white deliquescent solid, m.p.114°C. It is used in the manufacture of dyes and as a meat preser-vative.

Ammonium Bicarbonate or Ammonium Hydrogen Carbo-nate (NH_4HCO_3): A colourless crystalline's solid. It is used in baking powder.

Ammonium Bromide (NH_4Br): A colourless crystalline solid with a saline taste. It turns yellowish in air. It is obtained when HBr and NH_3 read in the gaseous state or solution.

Ammonium Carbonate [$(NH_4)_2$ CO_3]: A white solid, strongly smelling of ammonia due to its decomposition. It is fairly soluble in water. It obtained by heating a mixture of ammonium sulphate or chloride is o and chalk in iron retorts.

$$(NH_4)_2SO_4 + CaCO_3 \rightarrow CaSO_4 + (NH_4)_2CO_3$$

It is used as laboratory reagent in qualitative ana-lysis, smelling salt, and in dyeing. It is also used in baking powders.

Ammonium Chloride (NH_4Cl): A white granular, fibrous crystalline solid. It is fairly soluble in water and causes cooling in solution. It is prepared by boiling equimolecular quantities of ammonium sulphate and sodium chloride in solution

$$(NH_4)_2SO_4 + 2NaCl \rightarrow Na_2SO_4 + 2NH_4Cl$$

It is used as laboratory reagent. It is also used as flux in soldering, galvanising in medicine and calico printing.

Ammonium Chromate [$(NH_4)_2CrO_4$]: A golden yellow solid. It decomposes on heating to the dichromate, NH_3 and H_2O. It is soluble in water.

Ammonium Dichromate ($(NH_4)Cr_2O_7$): A red crysta-lline solid. It is highly soluble in water.

Ammonium Fluoroborate (NH_4BF_4): A colourless crystal. It is used as an electrolyte.

Ammonium Flouride (NH_4F): A white deliquescent solid. It is used as a disinfectant and preservative.

Ammonium Hydroxide ($NH_4\ OH$): An aqueous solution of ammonia. It is a weak base.

Ammonium iodide (NH_4I): A colourless salt. It is highly soluble in water. It is used in photographic industry.

Ammonium Molybdates: A range of salts obtained from ammonia and molybdic acid. The commercial salt is $(NH_4)\ 6M_0O_7O_{24}\ .4H_2O$.

Ammonium Nitrate (NH_4NO_3): A colourless crystalline salt, which deliquesces in moist air. The salt is highly soluble in water and also dissolves in alcohol.

Ammonium Nitrite (NH_4NO_2): A yellowish solid obtained by part oxidation of NH_3 or from $Ba(NO_2)_2$ and $(NH_4)_2\ SO_4$. It breaks up on heating to N_2 and H_2O.

Ammonium Perchlorate: A white crystallinesolid. It is used as an oxidizer in acid propellants.

Ammonium Phosphates [$(NH_4)HPO_4$ and $(NH_4)3PO_4$: Formed from NH_3 and phosphoric acid. They are used as fertilizers.

Ammonium Sodium Hydrogen Orthophosphate ($NH_4Na.HP\ O_44H_2O$): A white crystalline acid. It is used as a flux.

Ammonium Sulphate [$(NH_4)_2\ SO_4$: A white crystalline acid when heated to 300°C decomposes as :

$$(NH_4)_2SO_4 \rightarrow NH_4.HSO_4 + NH_3$$

It is used to prepare various other salts of ammo-nium and largely used as a fertilizer.

Ammonium Thiocyanate (NH_4SCN): A colourless crystalline substance, m.p. 149.6°C. It is used as a herbicide and in textile industry.

Ammonolysis: A chemical reaction in which one group of an organic compound is changed to an amine group, by reacting the compound with ammonia.

Ampere (Symbol A): The unit of current in the SI is the ampere. A current of IA is the flow of 1 coulomb, C of charge, is specified number of electrons, for one second past given point.

$$q = it$$

where q is charge, i is current, and t is time meas-ured respectively in coulombs amperes and seconds.

Amorphous: Some materials, such as, glass, glue, tar and asphalt, although they possess a certain degree of hardness and rigidity, are not characterized by any definite crystalline form. For this reason, they are classified as pseudo solids and are called amorphous material. When heated, they do not melt at fixed temperatures as do crystalline solids, but as the temperature is raised, they gradually soften and become more mobile.

Amphetamine ($C_6H_5CH_2CH\ NH_2CH_3$): A drug used in the form of the Sulphate which stimulates the central nervous system in cases of depression. Its trade name is Benzedrin.

Amphoteric: Metal hydroxides that are relatively insoluble in natural water but dissolve in either a strongly acidic or strongly basic medium. Examples include $A(OH)_3$, $Zn(OH)_2$, $Cr(OH)_3$ and $Sn(OH)_2$.

Amphoteric Oxide: An oxide that will neutralize either an acid or as a base.

Amyl Acetate or Pentyl Acetate or Banana Oil ($CH_3COO\ C_5H_{11}$): An ester c amyl alcohol and acetic acid. A colourless liquid b.p. 148°C with a odour of pear drops.

It is obtained by heating amyl alcohol with potassium ethanonat and sulphuric acid. It is used as solvent cellulose, acetate lacquers an paints. It is also used in perfumes and flavours.

Amyl Ether: A colourless liquid with a pleasant odour, b.p. 172.5 173°C.

It is obtained by heating amyl alcohol with sulphuric acid. It i used as a high boiling ether type solvent and as a solvent for fats an essential oils.

Amyl Alcohol ($C_5H_{11}OH$): A colourless liquid with a characteristic smell It exists in various isomeric forms. Commercial amyl alcohol is mainly isoamyl alcohol, b.p. 131.4°C. It is obtained from fused oil and used a a solvent.

Amylopectin: Two kinds of starch molecules occur in nature. About 1C to 20 percent consist of long unbranched chain molecules, and are called amylose. The other component, named amylopectin is a highly branched polymer in which most monomers are joined by 1, 4 linkage, with chain branching occurring through 1, 6 linkage. The linear polymer amylose is soluble in hot water whereas amylopectin is not. Both forms represent plant-energy storage which can be use directly by humans.

Amylase: An enzyme occurring widely in animals, plants and bacteria which reduces starch or other such complex sugars into glucose like simple Sugars.

Amylose: Two kinds of starch molecules occur in nature. About 13 to 20 percent of long unbranched chain molecules.

Anabolism: All the metabolic reactions that synthesis complex molecules from more simple molecules.

Anaesthetics: Compounds used to remove the perception of pain and other stimuli, e.g., N_2O, ethyl vinyl ether, cyclopropane, ethyl chloride and cocaine.

Anatase: A crystalline form of natural titanium oxide, TiO_2.

Analysis: The process of resolving a pure substance into two or more simpler constituent substances or, more commonly, decomposition.

Amino group: The group—NH_2.

Andalusite (Al_2SiO_3): A form of aluminium silicate. It is used for metallurgical furnace linings and as a refractory material.

Andrews' Experiment: An investigation into the relationship between pressure and valence for a mass of carbon dioxide at constant temperature. The resulting isothermals showed clearly the existence of a vertical point and led to greater understanding of the liquefaction of gases.

Andrews Titration: A titration for the estimation of reducing agents. The reducing agent is dissolved in concentrated hydrochloric acid and titrated with potassium iodate (V) solution. A drop of carbon tetrachloride is added to find out the end point. The colour of iodine will disappear. The iodate is reduced to ICI and reducing agent oxidized.

Anation: The removal of uncharged ligand, e.g., removal of H_2O from a complex with Cl^-.

Anethole (CH_3CH: $CHC_6CH_4OCH_3$): A white crystalline powder, m.p. 22.5°C. It is used in medicine, flavouring and in perfumes.

Angstrom (Å): A unit of length equivalent to 108 cm (1010 m).

Anhydrides: The basic oxides behave its though they were bases with the water removed. The acidic oxides behave as though they were acids with the water removed. They are often called anhydrides, of bases or acids

Anhydrite: Naturally occurring form of calcium sulphate ($CaSO_4$); generally used for drying organic solvents.

Anisaldehyde or p-Methoxy-Benzaldehyde: An oil (b), 521 K) which gives anisic acid on oxidation and anisyl alcohol on reduction.

Anhydrite: Naturally occurring form of calcium sulphate ($CaSO_4$); generally used for drying organic solvents.

Anionic resis: An lion exchange material that can exchange anions, such as Cl and OH for anions in the surrounding Medicine, such resins are used for a wide range of analytical and purification purposes.

Anisotropic: A true descriptive of certain substances which have one or more physical properties that differ according to direction. Most crystals are aminotropic.

Aniline: A colourless poisonous liquid with unpleasant odour, b.p. 184°C. It is practically insoluble in water but readily soluble in organic compounds.

Anion: A negatively charged ion formed by addition of an electron to atom or molecule. Monoatomic anions are named by dropping the ending name of the element and adding the ending -ide : H hydride ion, Cl chloride, N^{3-} nitride, S^{-2} sulphide, P^{-3} phosphide.

Anion exchanger: Polymers used in ion exchange reactions and chromatography; having amine or quarternary ammonium groups as integral part of the resin and an equivalent of anions such as Cl^-, Br, SO_4^{-2} or OH^- etc. groups. The anions of the ion exchange resin get exchanged by the ions (X^-) of the sample to be analysed or purified. The sample ions (X^-) and the exchanger can be regenerated by elution with a suitable base or electrolyte solution.

Aniline Hydrochloride ($C_6H_5N\ H_2HCl$): A colourless crysta-lline salt (m.p. 471 K) very soluble in water. It is precipitated by passing dry hydrogen chloride through a solution of aniline in ether. It is used in calico printing.

Anisaldehyde or Aubepine: A colourless oily liquid, the paraform of $CH_3OCC_6H_4\ CHO$, b.p. 247°C. It is used in cosmetics and perfumes.

Anisole or Methoxy Benzene ($C_7H_8O_7$): A colourless liquid, b.p. 155°C. It is obtained by the action of dimethyl sul-phate on a solution of phenol in excess of alkali. It is has an aromatic odour and is used in perfumes.

Anisotropy: A property exhibited by some materials some characteristics of the material have different values in different directions, e.g. refractive index.

Anisyl Alcohol or Anisalcohol ($CH_3OC_6H_5CH_2OH$): A colo-urless liquid. b.p. 258.8°C. It is used in perfumes.

Annealing: A process iii which the newly shaped articles are cooled slowly. If they are cooled quickly, they become brittle on account of the high internal strain.

Anodizing: A process of anodic oxidation of metals with the object of giving protection against corrosion.

Anode Mud: On passing the electric current, the pure metal generally dissolves from the anode and is deposited on the cathode. The soluble impurities go into the solution, while insoluble matter settles at the bottom and is called anode mud.

Anode Oxidation: A process at an anode during electrolysis in which a negative ion loses one or more electrons to the anode, *e.g.*, $2Cl^- \rightarrow Cl_2$ + 2e (Chlorine discharged).

Anomeric effect: Effort of the hydroxyl group, a polar aglycon group, at C—1 in a hexose molecule, that tends to assume the axial orientation rather than the equatorial. It is because the interaction between lone pairs of electrons of the ring-oxygen atom and those of the substituent is less for an axial rather, than for an equatorial substituent.

Antiacids: Substances used to reduce the amount of acid in stomach. They can he direct neutralizing ($NaHCO_3$), buffering (magnesium silicate, sodium silicate) or ion exchanging (zeolites).

Anthracene ($C_{14}H_{16}$): A colourless plate with a slight blue fluorescence, m.p. 217°C', b.p. 354-355°C, It is present in coal tar (1/2%) from which it is separated by fractional distillation. Of little use commerically except as a route to anthraquinone.

Anthranilic Acid or O-Aminobenzoic Acid C_6H_4-NH_2COOH): A white solid (m.p. 418 K) obtained by the reduction of o-nitrobenzoic acid. It is sweet in taste and soluble in water, alcohol and ether. It is used in the manufacture of indigotin and methyl anthranilate is used in perfumery.

Anthraquinone ($C_{14}H_8O_2$): Colourless needles or prisms which can be purified by sublimation or stream distillation. Its melting point is 285°C. It is used as a bird repellant.

Anthocyanins: The universal soluble violet, red and blue pigment of flowers, fruit and leaves.

Antichlor: Substance used to remove chlorine from materials after bleaching *e.g.*, sodium thiosulphate ($Na_2 S_2O_3$).

Antibiotics: Group of organic compounds secreted by fungi and capable at low concentration of inhibiting the growth of or destroying other microorganisms like bacteria.

Antichlor: A substance used to remove chlorine from materials after bleaching, *e.g.*, sodium thiosulphate, $Na_2S_2O_3$.

Antiferromagnetism: Positive magnetic behaviour in which the magnetic susceptibility drops with decreasing temperature.

Anti-Foaming Agents: Compound which prevent the formation of foams, *e.g.*, polyamides.

Antiknock: A substance that helps to control the burning rate of gasoline. The most widely used substances for this purpose are tetra ethyl lead $(CH_3)_4Pb$ and tetramethyl lead. The premium gasolines contains 2 or 3ml of one of these lead compounds per, gallon with a increase of 15 in octane rating.

Antimatter: Matter formed of antiparticles. It is that for every fundamental particle of matter there is an antiparticle, of the same mass but opposite charge.

Antimetabolites: The compounds derived from metabolites of essential foods (proteins, vitamins, fats or carbohydrates) by a minimum structural change, mostly by isosteric replacement in their structure *e.g.*, 5-fluorouracil, 5-bromouracil, 5-fluoroorotic acid and 2, 6 diaminopurine are nucleobase antimetabolities. An antimetabolite, owing to the structural resemblance, can replace the corresponding natural metabolite as a forcing building unit in biopolymers and interfere with normal insect growth.

Antimony: (Sb) A lustrous metallic element belonging to group 15 and a poor conductor of electricity. The various allotropic forms are yellow antimony (a-anti-mony), explosive antimony and metallic antimony (b-antimony). An explosive antimony is obtained by electrolysis of a solution of antimony trichloride in hydrochloric acid. It explodes on rubbing or scratching accompanied by the evolution of dense white fumes of $SbCl_3$. It is mainly used in the manufacture of alloys to which it adds hardness and resistance to corrosion.

Antimony (III) Chloride ($SbCl_3$): A white soft solid. It is very deliquescent and is hydrolysed by water to give a precipitate of the oxychloride (SbOCl)

$$SbCl_3 + H_2O \rightarrow SbOCl^- + 2HCl$$

It is used as a caustic in medicine, for etching of metals, or for the cleaning of gun barrels.

Antimony (V) Chloride (Antimony Penta Chloride) ($SbCl_5$): A colourless liquid prepared by treating the molten trichloride with chlorine. It is mainly used in organic chemistry as a strong chlorinating agent.

Antimony (III) Fluoride (SbF_3): A solid having m.p. 292°C. It is mainly used as mild fluorinating agent.

Antimony Hydride or Stibine (SbH_3): An unstable colour-less gas, m.p. 880°C, b.p. 170°C. It decomposes easily to antimony.

Antimony Iodide (SbI_3): A red or yellow solid, m.p. 170°C.

Antimony (III) Oxide (Sb_2O_3): A white powder. It melts at 650°C. It is soluble in water, but behaves as amphoteric. It is used in medicine and for the preparation of tartaremetic.

Antimony (V) Oxide (Sb_2O_5): A yellow solid, insoluble in water. It is prepared by heating antimony with concentrated nitric acid or by the hydrolysis of antimony penta chloride. It is a precipitates iodine from an acid solution of potassium iodide

$$Sb_2O_5 + 10HCl \rightarrow SbCl_5 + 5H_2O$$

$$SbCl_5 + 2KI \rightarrow SbCl_3 + 2KCl + I_2$$

It is used for the preparation of antimonates.

Antimony Pentasulphide (Sb_2S_5): A yellow insoluble powder. It is used in the vulcanisation of rubber and as a pigment. Antimony Potassium Tartrate or Tartar Emetic: A white soluble poisonous powder. It is used as an emetic and as a mordant.

Antimony Sulphate [$Sb_2(SO_4)_3$]: A white crystalline insoluble solid. It is used in explosives.

Antioxidants: Materials used to prevent harmful effects of oxidation in rubber, plastics and paints.

Antimony Trisulphide or Stibnite (Sb_2S_3): A black or red insoluble crystalline solid, m.p. 55°C. It is used in fireworks, pigments and matches.

Antiparticle: An elementary particle with charge properties opposite to those normally found, *e.g.,* positive electron.

Antipyretics: Drugs used to decrease temperature in case of fever *e.g.* salicylates.

Antiseptics: Substances which stop the growth of microorganisms, generally applied to living tissue, *e.g.,* I_2.

Apatite [$Ca_5(PO_4)3F$]: A phosphate mineral having fluoride and phosphate anions. P is used for the production of phosphorus and its compounds including super phosphate.

Apomorphine ($C_{17}H_{17}NO_2$): A crystalline alkaloid derived from morphine.

Apophysis: A swelling or expansion found in the organ-Z, or Z-organ, in species of *Xiphinema.*

Aqua Regia: A solution containing hydrochloric acid and nitric acid in the ratio of three moles of the former to one of the latter. It dissolves noble metals like platinum and gold which are unaffected by nitric acid alone.

Aqua Ions: A hydrated metal ion in aqueous solution. It can also be applied to hydrated ions in complexes, e.g., $[CO(H_2O)_6{}^{+2}]$.

Aquation: The process of replacement of other ligands by water, e.g., $[CoNH_3)_4\ .(H_2O)(NO_3)]^{2+} \rightarrow [CoNH_3)_4\ (H_2O_2)_2]^{3+}$

Aqueous (Aq.): Substance dissolved in water or contai-ning enough water aqueous acetic acid etc., Aq. is the symbol used to denote water *e.g.*, aq, HCl means hydrochloric acid in water.

Arachidic Acid or Archic Acid [$CH_3(CH_2)COOH$]: A white crystalline insoluble oil. It is used insoluble solid, m.p. 76.3°C. It is obtained from peanut 1 in lubricants, waxes and plastics.

Arene: An organic compound containing a benzene ring, *i.e.*, an ornamatic hydrocarbon or derivative.

Argon: A monoatomic inert noble gas abundant in the atmosphere. Air has 0.94% argon. Argon is obtained from liquid air by fractional distillation. It is used to fill incandescent light bulbs. The chemical inertness of argon minimizes deterioration of the filament by oxidation. It is also used as an inert atmosphere in welding and high temperature metallurgical processes.

Symbol : Ar; a.n. 18; m.p. –189.2°C p. –185.7°C; r.d. 1.78 kg m^{-3}.

Artificial Radioactivity: The radioactivity of elements which have been made radioactive by bombardment with particles of high energy.

Arndt-Eistert Synthesis: A procedure for converting a carboxylic acid to its next higher homologue or to a derivative of a homologous acid, *e.g.*, ester or amide.

Aromatic Compounds: A class of organic compounds whose structure is based on that of benzene. They contain ring-like structures made of six carbon atoms.

Aromaticity: The property of extra stability and inertness shown by unsaturated cyclic organic compounds which obey Huckel rule stating that there should he a planar ring structure with a total $(4n + 2)$p - electrons, where n is a natural number. Compounds containing benzene ring are called benzenoids while those having no benzene ring are called nonbenzenoid aromatics.

Arrangement of Electron in Orbits: In 1912 Bohr and Bury simultaneously put forward two identical schemes for the arrangement of extranuclear electron. Now it is called Bohr-Bury scheme. It may be summarized as follows

1. The maximum number of electrons which each orbit can contain is $2n^2$ whose n gives the number of the orbit.
2. The maximum number of electrons in the outer most orbit is 8 and in orbit next to orbit most is 18.
3. It is not necessary for an orbit to be complete before another Commences to be formed.
4. The outer most orbit cannot have more than 2 electrons and orbit next to the outer most cannot have more than nine, so long as the next inner orbit in each case has not received the maximum number of electrons as required by rule (1).

Arrhenius Equation: An equation relating the rate constant for a reaction to the product of a constant (the frequency factor) and an exponential term containing the activation energy; $K = Ae^{Ea/RT}$. In this equation, e is the base of the natural logari-thm scale. The term Ea is the activation energy; R is the gas constant and T is a absolute temperature; A is a constant or nearly so, as temperature is varied. It is called frequency factor.

Arsenic: Poisonous metallic element belonging to group 15. Three crystalline isotopes of arsenic are known viz., yellow arsenic (a-arsenic), black arsenic (b-arsenic) and grey arsenic (g-arsenic). Compounds of arsenic are used as weed killers and insecticides in agriculture. Small amount of arsenic (0.10.2%) are added to lead for making lead shots for guns. Addition of arsenic to metals makes them hard.

Arsenic Acid ($H_3ASO_4.H_2O$): A while soluble crystalline powder m.p. 35.5°C. It is used in the manufacture of arsenates.

Arsenic (III) Chloride (Arse-nious chloride $ASC1_3$): A poisonous only liquid. It is fumes in moist air due to hydrolysis with water vapour

$$ASCl^3 + 3H_2O \rightarrow AS_2O_3 + 6HCl$$

Arsenic (III) chloride is covalent and exhibits nonmetallic properties.

Arsenic (III) Oxide (AS_2O_3): A glassy amorphous powder which soon becomes opaque and reverts to a crystalline form. It readily dissolves in hot water and the solution is weakly acidic due to the formation of arsenious acid : H_3AsO_3 or $HAsO_2$. It is used insecticide and weed killer. It is also used for making free arsenic.

Arsenic (V) Oxide (AS_2O_5): A white solid which readily dissolves in water to form a solution of arsenic acid

$$AS_2O_5 + H_2O = H_3AsO_4$$

It can be prepared by heating arsenious oxide with oxygen at 300°C and under high pressure of atmosphere. It is an oxidising agent and liberates chlorine from hydrogen chloride.

Artificial Radioactivity: Radioactivity induced by bombarding stable nuclear with particles of high energy. For example

$$^{27}_{13}Al + {}^{1}_{0}n \; {}^{24}_{11}Na \; {}^{4}_{2}He$$

represents the bombardment of aluniviniome with nucleus to produce an isotopes of dsokium *e*. All the bransuranic elements (atomic numbers 93 and above) are artificially ferrogative since they do not occur in nature.

Aryl Group: An organic groups derived by removing a hydrogen atom from an aromatic hydrocarbon or derivative.

Arsenious Sulphide: AS_2S_3 is used as a yellow pigment for oil paints.

Arsine (AsH3): A colourless with a garlic odour. It is insoluble in Water. It is usually prepared in the laboratory by the action of metal and dilute acids on arsenous oxide. It is used in the detection of arsenic

Artificial Radioactivity: The radioactivity of elements which have been made radioactive by bombardment with particles of high energy.

Asphalt: A black semisolid sticky substance which is composed of bitumen with mineral matter. It is used in road-making and building.

Association: The combination of molecule of a substance with those of another to form more complex species. An example is a mixture of water and ethanol (which are termed associated liquids) the molecule of which combine via hydrogen bonding.

Astatine (At): A solid with properties more metallic than iodine. It dissolve in nonpolar solvents like carbon disulphide and carbon tetra chloride, It is highly radioactive with a short half-life period. The most common isotopes has a half-life period of 8.3 hours. It can be made by bombarding bismuth nuclei with alpha particles.

Asymmetric Carbon Atom: A carbon atom to which four different type of groups are bonded.

Atom: The smallest part of an element that has the same properties as the bulk. Dalton considered atom as all indivisible and structureless particle. Later workers showed experimentally that each atom consists of a dense positively charged nucleus made up of neutrons and protons. Negatively charged electrons orbit at different of energy levels called electron shells surrounding the nucleus.

Atneolysis: The separation of gases by using their different rates of diffusion.

Atmosphere: A unit of pressure defined as 101325 pascals (atmospheric). The atmosphere is used in chemistry only for rough values of pressure; in particular stating the pressures of high pressure industrial processes.

Atomicity: The number of atoms in a molecule of an element or a compound, when the element or compound consists of separate molecules, e.g., Neon has an atomicity of 1, hydrogen of 2 and carbon dioxide of 3.

Atomic absorption spectro-photometry (AAS): An analy-tical technique for estimation of elements and their compounds by measuring the excitation energy for the atoms. If a radiation of suitable wavelength is passed (using a suitable hollow cathode lamp) through a flame containing sample atoms (Ag, AI, B, Ba Be, Bl, Ca, Cd, Co, Cs, Fe, Hg, K, Li, Mg, Mn, Ni, Sb, Se, Si, Sn, Ti and Zn), a fraction of the atoms will absorb the radiation and the fall in the intensity of incident radiation will depend upon the concentration of the sample atoms in the flame. The method follows lambert-Beer's law of, absorption spectrometry and even very dilute solutions can be analysed using longer path length. The method is a handy technique for determination of micronutrients (Zn, Fe, Mn, Mo, Ni and Co) and traces of polutants like Hg in river water.

Atomic Constants: Electronic charge, e = 4.89298 x 10^{-10} e.s.u.; electronic rest mass, *m* = 0.91091 x 10^{-27} g; velocity of light C = 2.9979 ´ 10^{-8} m/sec; Planck's constant, h = 6.6256 x 10^{-10} erg/sec.

Atomic mass unit (a.m.u.): Unit used to express mass of individual isotope of elements. It is equal to the mass of a proton approx., 1.66 ´ 10^{-24} g.

Atomic Number (z): The number of protons in the nucleus of an atom.

Atomic number = Mass number – Number of neutrons.

Atomic Radius: The distance between the nucleus and outer boundary of the electrons. Since a free uncombined atom has sharp boundary to its mobile electrons (cloud) it is not possible to measure atomic radii for individual atom. The values, however, are derived indirectly from its internuclear distances. Half the internuclear distance is taken to be the atomic radius of the atom.

Atomic Volume: The volume occupied by 1 gram atom of an element.

Atomic Weight: The ratio of the average mars of an atom of an element to 1/12 of the mars of an atom of carbon isotope. It is also called relative atomic weight. It is the combined weight of proton and neutron in the nucleus.

Atto (a): A prefix denoting 10^{-18}. For example,

I attometer (am) = 10^{-18} meter (m).

Atropine ($C_{17}H_{23}NO_3$): A colourless crystalline insoluble alkaloid m.p. 115°C. It is very poisonous and has a powerful effect upon the nervous system. It is used in medicine to dilate the pupil of eye.

Aufbau Principle: The electronic configuration of lowest energy is the one in which each electron is placed in the lowest energy hydrogen like orbital that is available.

Auric Bromide: Gold (III) Bromide.

Auric Chloride or Gold Chloride ($AuCl_{13}$): A red soluble crystalline solid. It is used in gliding glass and Photography.

Aurin or Resolic acid: A red crystalline solid obtained by diazotising rosaniline and boiling the diazonium com-pound so obtained with water. It finds use as an indicator and printing wall papers.

Aurous Ion (Au^+): Au^+ is an oxidizing agent powerful enough to oxidize water. Au^+ ion is also unstable with respect to disproportionation. It does not exist as a simple cation in aqueous solution.

Auto Catalysis: A process which takes place when one of the products of a chemical reaction acts as a catalyst for the reaction, e.g., potassium permanganate in acid solution oxidizes ethanedioic acid (oxalic acid). The Mn^{2+} ions, a product of the reaction, catalyzed the reaction.

Autoclave: A pressure-cooker like apparatus used to sterilize materials with steam under pressure. It is also used for reaction between gases under pressure and liquid or solid reactants.

Auxins: A group of plant hormones (both natural and synthetic) which when sprayed in low concentration promote the growth of the plant e.g., b-Indoleacetic acid. Physiological effects of auxins include breaking or inducing of seed dormancy, roofing of cutting, cell enlargement, apical dormance and induction of partenocarpy, inhibition of abscission and control of synchronous senescence.

Avogadro's Law: Equal volume of all gases at the same temperature and pressure contain the same number of molecules. Two very useful applications of Avogadro's Law are:

(1) The formula of the gas can be derived from the reacting volumes.

(2) The relative molecular mass of a gas is equal to 2X vapour density.

Avogadros' Number (L or N): The number of particles (atoms or molecules) in one mole of any pure substances

$$L = 6.023 \times 10^{23}$$

Azetidine: A colourless liquid, b.p. 61°C, smelling of ammonia. It dissolves in polar solvents.

Azides: A hydrazoic acid reacts with electropositive metals forming salts called azides, but unlike other acid + metal reaction, no hydrogen is evolved.

Azo Benzene ($C_{12}H_{10}N_2$): An orangered crystal, m.p. 68°C. It is the simplest compound having N = N chromophore.

Azo Compound: A type of organic of the general formula RN : NR; where R are aromatic groups. Azo compounds can be formed by coupling a diazonium compound with an aromatic phenyl or amine. Most are coloured because of the -N : N- (azo) group.

B

Babbit Metal: A type of alloys with a high proportion of tin and small amount of copper and antimony. It is used for bearings.

Babo's Law: The addition of a nonvolatile solid to a liquid in which it is soluble lowers vapour pressure of the solvent in proportion to the amount of substance dissolved.

Back Bonding: Overlap of a filled orbital, commonly *a d* orbital, of an acceptor with a vacant orbital *d* or *p* (Bonding or Antibonding) of a ligand, e.g., metal carbonyes.

Baddeleyite: A kind of ZrO_2 and the main mineral source of Zr.

Balance: An apparatus used to weigh solids and liquids contained in vessels.

Balmer Series: The visible spectrum of hydrogen. It consists of a series of sharp distinct lines, the wavelength, X of which may be represented by the formula :

$$1/\lambda R(1/2^2 - 1/n^2)$$

n =3, 4, 5 etc.; R = Rydbrog's constant.

Band Spectra: An emission or absorption spectrum comprised of a number of fluted bonds each having one sharp edge, Each band consists of large number of closely spaced lines.

Banana Bond (Bent Bond): In standing compounds the analyse assured on the basis of hybridization are after not equal to the analyse obtained by going the atomic centres. In there cases, it is assumed that the bonding orbital is burnt or banana like in shape cyclopropane is an example, in which geometric consideration imply a bond analyse of 60° while sp^3 hydrization implies an interorbital analyse of around 100°, giving a banana bond.

Bar: A unit pressure defined as 105 pascals. The milibar (mb) is more convenient. It is used for measuring atmospheric pressure in meteorology.

Barium Hydrogen Carbonate (Barium Bicarbonate Ba $(HCO_3)_2$: A compound that occurs only in aqueous solution. It is formed by the action of cold water containing carbon dioxide on barium carbonate, to which it reverts on heating :

$$BaCO_3 + CO_2 + H_2O = B(HCO_3)_2$$

Barbitone or Diethylmalonylurea ($C_8H_{12}N_2O_3$): A white crystalline solid having m.p. 191°C. Its sodium salt is used as sedative.

Barbituric Acid or Malonylurea ($C_4H_4N_2O_3$): A colourless crystalline solid obtained by heating malonic ester with urea and sodium ethaoxide. It is a dibasic acid.

Barff Process: A process of preventing iron from rust by the action of steam on the redhot metal, resulting in a surface coating of black oxide of iron, Fe_3O_4.

Barium: A dense, lowmelting reactive metal. It belongs to II A group of the periodic table. It occurs as carbonates, $BaCO_3$ in witherite and more, widely as sulphate, $BaSO_4$ in Barytes. It catches fire spontaneously on exposure to air in powder form. It is malleable at 850°C Barium compounds give a green colour to flame which is used in qualitative analysis.

Symbol Ba, a.n. 56; m.p. 714°C; b.p. 1537°C; r.d. 3.51.

Barium Carbonate ($BaCO_3$): A white insoluble salt that occurs in nature as the mineral witherite. It is used as a constituent of rat poisons.

Barium Chloride ($BaCl_2$): A white solid obtained by heating barytes ($BaSO_4$) with $CaCl_2$ and carbon. It crystallizes out as $BaC1_2.2H_2O$. It is soluble in water and nonhygroscopic. It is used as laboratory reagent

Barium Chromate ($BaCrO_4$): A yellow pigment which is insoluble in water.

Barium Hydroxide [$Ba(OH)_2$]: A white solid formed by heating barium carbonate in super heated steam, It crystallizes from water as the octahydrate [$Ba(OH)_2.8H_2O$]. It is used for titrating weak acids.

Barium Oxide (BaO): A white powder which slakes with water with evolution of heat. It is used in the preparation of oxygen from (Brin's process).

Barium Peroxide (BaO_2): A white powder obtained by heating barium oxide in oxygen.

Barium Sulphate ($BaSO_4$): A fine white powder that occurs in nature as mineral barytes. It is exceedingly insoluble in water. It is used in the manufacture of coloured paints and in the compounding of rubber. Patients are made to swallow a suspension of $BaSO_4$ before taking X-ray photograph of their digestive system, since it is opaque to these rays. It is a source of other barium compounds.

Barium Titanate ($BaTiO_3$): A solid state compound obtained from $BaCO_3$ and TiO_2 at high temperature. It is used in transducers, ultrasonics and for polishing.

Barometer: A device for measuring atmospheric pressure in terms of the height of a liquid column sustained by that pressure.

Barylium ($BeCO_3$): An unstable solid prepared by prolonged treatment of a suspension of berylium hydroxide with carbon dioxide.

Baryon: Any member of the group of elementary particles called hyperons or nucleons. Protons and neutrons are example of baryons.

Barytes: A mineral form of barium sulphate ($BaSO_4$).

Basalt: A rock of 4 volcanic origin, chemically resembling feldspar.

Base: Any ion or molecule that can receive hydrogen ion or proton or donate an electron. The hydroxide ion (^-OH) is also called a base. All alkalis like sodium hydroxide are bases.

Base: A compound that release hydroxyl ions, OH; in aqueous solution, Basic solutions hence a p^H greater than 7. In the howryBrounsted treatment a base is a substance that tends to accept a proton. This OH is basic as it accepts H^+ to form water, but $2H_2O$ is also a base (although some what weaker) because it can accept a further proton to form H_3O^+. In this treatment the ions of classical mineral acids $(SO_4)_2$ and NO_3 are weak conjugate bases of their respective acids. Electron pair donors, such as trimely amine and pyridine, are example of organic bases.

Base Catalysed Reaction: The added reaction and Claisen condensation is done in the presence of base, in which the first step is abstraction of a proton to give a carbonion.

Base Metal: A metal like iron, lead or copper, distinguished from a valuable metal, such as, gold or silver.

Basic: The tendency of a substance to give OH ions. The pH of the solution is greater than 7.

Basic Oxide: The metal oxides give basic solution when they are dissolved in water, e.g., NaO_2, K_2O

$$Na_2O + H_2 \rightarrow 2NaOH\ (aq)$$

Metal oxide (1) base

Base Unit: A unit that is defined in terms of reproducible physical a base unit in the SI, being defined in terms of the wavelength of a particular light emission.

Basic Salt: A compound intermediate between a normal salt and hydroxide or oxide. The terms is often restricted to hydroxy halides [such as Pb(OH)Cl, $Mg_2OH_3Cl_2.4H_2O$, and Zn(OH)F] and hydroxy salts.

Batch Process: A manufacturing process in which the reactants are fed into the process in fixed quantities (batches), rather than in a continuous flow. At any particular instant all the material, from its propagation to other final product, has reached a definite stage in the process. Banking a coke is an example of a batch process.

Bath Salts: The main constituent is generally sodium sesquicarbonate, $Na_2CO_3.Na\ HCO_3.2H_2O$ or some other soluble sodium salt to soften the water.

Bauxite: The ore of aluminium which is essentially Al_2O_3 with additional water molecules. The ore is contaminated iron oxide and silica.

Beckmann Thermometer: A type of the thermometer designed to measure differences in temperature rather than scale degrees. Beckmann thermometer have a larger bulb than convion thermometers and a stem with a small internal diameter.

Belleslein Test: A method for detecting the presence of halogen in an organic compound. A piece of copper gauze is heated in an oxidizing flame until the flame is no longe tinged green; the compound is placed on the gauze which is reheated.

Benedict Solution: Aqueous solution of Na_2CO_3, $CuSO_4$ and sodium citrate. It gives red-yellow colour or precipitates.

Beneficiation: The separation of ores into valuable components and wastes can be achieved by various methods, *e.g.,* flotation.

Benzemide or Benzene carboxamide ($C_6H_5CONH_2$): A white crystalline solid (m.p. 403 K) obtained by the action of concentrated aqueous ammonia on benzoyl chloride,

$$C_6H_5COCl + 2NH_3 \rightarrow C_6H_5CONH_2 + NH_4Cl$$

It is insoluble in cold water but dissolves in hot water.

Benzene (C_6H_6): A colourless liquid, m.p. 5.5°C, b.p. 80°C with a peculiar smell. It is obtained by passing acetylene gas through a red-hot tube, It is inflammable, burning with a smoky flame. It is used as solvent for fats, resins, as motor fuel, and in drycleaning.

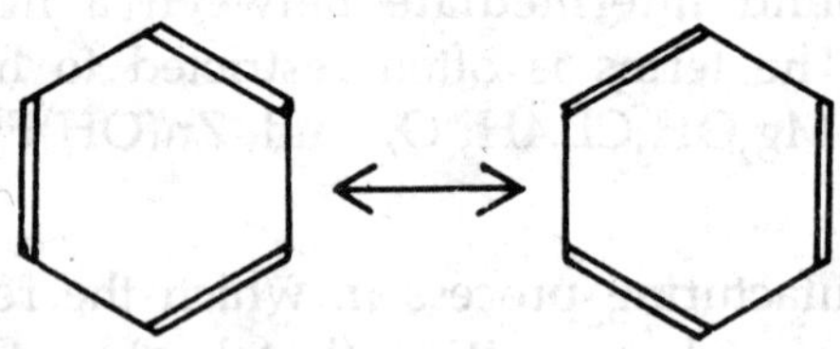

Kekule structures

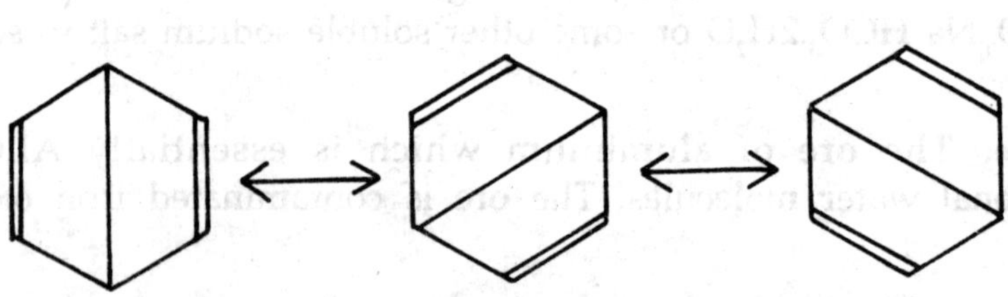

Dewar structure

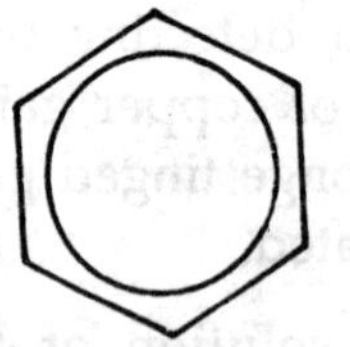

Representation

Benzene Carboxylic Acid (Benzoic Acid, C_6H_5COOH): White, pearly flakes (m.p. 395°C), sparingly soluble in cold water but readily soluble in hot water, alcohol and ether. It is obtained by the hydrolysis of benzylidyne chloride with milk of lime in the presence of iron powder (Catalyst). It is used as medicine and its salt sodium benzoate finds use as food preservative.

Benzene Carboxyl Chloride (Benzyl Chloride): A colourless fuming liquid, b.p. 197°C irritating odour. It is used to characterise the alcohol and as a benzoylating agent.

Benzene Carbaldehyde (Benzaldehyde, (C_6H_5.CHO): A colourless liquid, b.p. 179°C with a smell of bitter almonds. It is prepared in the laboratory by oxidising toluene with chromium trioxide in acetic anhydride. It is used in perfumery, in the dye industry and for flavouring purpose.

1, 2-Benzenediol or Catechol: A colourless solid, soluble in water, alcohol and other. It condenses with a number of compounds, like phthalic anhydride in the presence of sulphuric acid to form alizarin. It is used as a developer in photography and as an antioxidant in gasoline.

1, 3-Benzenediol or Resorcinol: A colourless crystalline solid (m.p. 383 K), very soluble in water, ethanol and ether. It couples with diazonium salts to form azo dyes. It is used for making dyes and as an antiseptic.

1, 4-Benzenediol or Quinol: A colourless crystalline solid (m.p. 443 K) obtained by reducing p-benzenoquinone with sulphurous acid. It is very soluble in water, alcohol and ether but insoluble in benzene. It is used as a photographic developer and as an antiseptic.

1, 2, 3-Benzenetriol or Pyrogallol: A colourless Crystalline solid (m.p. 485 K) obtained by heating solid gallic acid in a current of CO_2. It is soluble in water, alcohol and ether. It is used as a photographic developer, as hair dye and in antiseptic ointments.

1, 2, 4-Benzenetriol or Hydroxyquinol: A colourless crystalline solid (m.p. 413 K) obtained by the alkaline fusion of 1, 4 benzenediol in air. It is very soluble in water and the aqueous solution gives greenish brown colour with $FeCl_3$.

1, 3, 5-Benzenetriol or Phloro-glucinol: A colourless solid (m.p. 491 K) obtained by fusing 1, 3-benzenediol with NaOH in air. It is fairly soluble in water. Its aqueous solution tastes sweet and gives a bluish violet colour with $FeCl_3$. It gives red colour with carbohydrate in the presence of sulphuric acid and is used in the detection of carbohydrate.

Benzenediazonium Chloride ($C_6H_5N_2Cl$): A colourless crystalline solid, readily soluble in water, It turns brown on exposure to air. It is used in the preparation of aryl fluorides and iodides which cannot be obtained by direct halogenation. It is also used to get phenols, hydrocarbons, ethers, cyanides, amines, acids, nitroderivative and azo dyes.

Benzene Sulphonic Acid ($C_6H_5SO_2OH$): A white crystalline acid prepared by heating benzene with concentrated sulphuric acid at 350 K.

Benzenesulphonyl Chloride ($C_6H_5SO_2Cl$): An oily liquid, (b.p. 519 K with decomposition) soluble in ether. It is obtained by treating benzene with chlorosulphonic acid (excess).

$$C_6H_6 + 2ClSO_3H \rightarrow C_6H_5SO_2Cl + H_2SO_4 + HCl$$

Benzidine ($NH_2C_6H_4C_6H_4NH_2$): An aromatic base having, b.p. 128°C. It is used in dyestuff industry.

Benzil ($C_{14}H_{10}O_2$): A solid having, m.p. 95°C. It is obtained by the oxidation of benzoin with nitric acid.

Benzoic Anhydride [$(C_6H_5\ CO)_2O$]: A white crystalline substance, (m.p. 315 K) obtained by heating sodium benzoate with benzoyl chloride. It can be used in place of benzoyl chloride in benzoylation.

Benzoin ($C_6H_5CHOH.CO.\ C_6H_5$): (1) A crystalline, optically active substance, m.p. 133.7°C. It is used in organic synthesis. (2) Gum Benzamin. A natural brown aromatic resin formed from a certain tree. It is used in the manufacture of perfumes and cosmetics.

Benzonitrite or Phenyl Cyanide (C_6H_5CN): A colourless poisonous liquid, b.p. 190.7°C. It is used in organic synthesis.

Benzophenone or Dilphenyl Ketone ($C_6H_5COC_6H_5$): A colourless, pleasant smelling solid obtained by heating calcium benzoate

$$(C_6H_5COO)_2Ca \rightarrow C_6H_5COC_6H_5 + CaCO_3$$

It is soluble in alcohol and ether but insoluble in water,

Berzopyrene ($C_{20}H_{12}$): A yellow crystalline polycyclic hydrocarbon, m.p. 179°C. It occurs in small quantities in coaltar.

p-Benzoquinone or p-Quinone: A yellow prism (m. p. 389 K) obtained by oxidation of aniline with potassium dichromate and sulphuric acid. It is a source of hydroquinone and some sulphur dyes.

Benzylamine ($C_6H_5CH_2NH_2$): A colourless liquid (b.p. 458 K) having a characteristic ammoniacal odour resembling aliphatic amine.

Benzylidene Chloride or Benzyl Chloride ($C_6H_5CHC_{12}$): A colourless liquid, b.p. 205.2°C. It is used in the manufacture of dyes.

Benzoyl Peroxide [$(C_6H_5CO)_2O_2$]: A crystalline explosive substance, m.p.

106.8°C. It is used in bleaching flour and oils, and as a catalyst.

Benzyl group: The univalent radical $C_6H_5CH_2$.

Berkelium: A transuranic element belonging to group 3. Its most stable isotope $^{247}_{97}Bk$ has a half life period of about 1400 years.

Symbol : Bk, N = 97.

Beryllium: A grey light metal belonging to group 2, obtained by the electrolysis of a mixture of beryllium chloride and sodium chloride. It is used for making alloys, as moderator in nuclear reactors, as a source of neutrons and its sheets are used for windows of the X-ray tubes.

Symbol : Be ; a.n. 4 ; m.p. 1280°C ; b.p. 1500°C ; r.d. 1.85.

Beryllium Chloride ($BeCl_2$): A white crystalline solid prepared by burning the beryllium in an atmosphere of chlorine. It dissolves in water with the evolution of heat. The anhydrous salt is used as a catalyst.

Beryllium Flouride (BeF_2): Obtained by heating the double salt $BeF_2.2NH_4F$ in the absence of air. It is hygroscopic substance easily soluble in water.

Beryllium Hydroxide [$Be(OH)_2$]: A white gelatinous precipitate obtained by adding ammonia to the solution of beryllium salt. It is soluble in acids and alkalies to form beryllates, *i.e.,* Na_2BeO_2.

Beryllium Nitrate [$Be(NO_3)_2$ $3H_2O$]: A readily available salt obtained by the reaction between beryllium carbonate and nitric acid. It is used impregnating incandescent gas mantles.

Beryllium Oxide (BeO): A white powder insoluble in water. It is amphoteric in nature and is used as a refractory material.

Beryllium Sulphate ($BeSO_4$): Obtained by die action of hot concentrated sulphuric acid on beryllium oxide. On cooling the crystals of $BeSO_4.H_2O$ are formed. It hydrolysis in aqueous solution to form the hydroxide.

$$BeSO_4 + 2H_2O \rightarrow Be(OH)_2 + H_2SO_4$$

b:b-Dichloroethyl Ether [$(CH_2ClCH_2)_2O$]: A colourless liquid with chloroform like odour (b.p. 37 K). It is obtained by dehydrating ethylene chlorohydrin with sulphuric acid at 370 K. It is an excellent solvent.

Betaine or Trimethylglycine ($C_5H_{11}NO_2$): It is soluble in water and alcohol.

It is found in beets.

Beta Iron: An allotropic form of pure iron, It is stable between 768°C and 910°C and nonmagnetic.

Beta (b) Particles: The electrons emitted from the nuclei of certain radioactive atoms.

Beta Particle: An electron emitted by a radioactive substance as it decay.

Betaines: Trialkyl derivatives of amino acids e.g., $(CH_3)_3N^+$ CH_2COO^-. They exist as dipolar ions.

b-Phenyl Ethyl Alcohol or 2-Phenyl Ethanol ($C_6H_5CH_2C$ H_2OH): A colourless, sweet smelling oily liquid (b.p. 295 K) obtained by reducing phenyl acetic ester with sodium alcohol. It is used in perfumery.

Bicarbonate: A traditional name for a hydrocarbonate.

Bidentate: A legand that coordinate through two bonds from different parts of the molecule or anion, *e.g.,* carbonate, oxalate and ethylenediamine (NH_2CH_2 CH_2NH_2).

Bimolecular Reaction: A chemical reaction in which two molecules or other species react together, e.g., H2 + I_2 ® 2HI.

Binary Acid: An acid with the general formula HnX, where n is a nonmetal.

Binary compound: A chemical compound of two elements only that is denoted by the sulphide *e.g.,* calcium carbide, CaC_2.

Binding Energy: The energy needed to pull the nucleons of the nucleus apart or the energy released by the hypothetical formation of the nucleus by the condensation of the individual nucleons. It may be calculated from the difference between the sum of the masses of the constituent nucleons and the mass of the corresponding nucleus. For $^{35}_{17}$HCl this mass difference is 0.3204 and, therefore,

$$0.3204 \times 931 \text{ Me v/u} = 298 \text{ MeV}$$

The amount of the binding energy of a given nucleus tells the stability of that nucleus toward radioactive decay.

Biochemistry: The study of the substances found in living organism, including the chemical reactions occurring in such system.

Biosynthesis: The reactions by which organise his obtain the various compound needed for life.

Birkland-Eyde Process: Atmospheric nitrogen and oxygen can be made to combine under the influence of an electric are (temperature 3000-3500°C) to give nitric oxide. The nitric oxide is, oxidised to nitrogen dioxide which is then absorbed in water to give nitric acid.

Biphenyl or Diphenyl (C_6H_5 C_6H_5): An insoluble Colourless powder, b.p. 70°C, it is usual in the manufacture of dyes and in organic synthesis.

2, 2-bis (4-Chlorophenyl)-1, 1 trichloro ethane (DDT) or p, p¢-dichlorodiphenyl trichlorethane: It is synthesized by condensation of 1 mole of chloral with 2 moles of monochlorobenzene in the presence of sulphuric acid. It is practically insoluble in water, moderately soluble in polar solvent and in petroleum oils. It is used as an insecticide.

Bismarck Brown or Basic Brown: A basic azo-dyestuff to dye wool. It is used as hair dye and also used for cotton with tanin as mordant.

Bismuth: A silvery white metal belonging to group 15, The important Sources of the element are bismuth glance, Bi_2S bismuth telluride, Bi_3Te_3 and the oxide, Bi_2O_3 known as, bismuth ochore. It is used for making alloys of low melting points.

Symbol : Bi, a.n. 83; m.p. 271°C b.p. 1560°C; r.d 9.75.

Bismuth Carbonate [Bi2(CO_3)$_3$]: A white precipitate obtained by adding ammonium carbonate to a solution of bismuth nitrate. On heating it gives bismuth trioxide.

It is used in medicine for intestinal disorders and in connection with the X-ray examination of oesophagus and stomach for ulcers etc.

Bismuth Dichloride ($BiCl_2$): Obtained by heating bismuth with Calomel at 523 K.

Bismuth Monoxide (BiO): A black powder obtained by heating basic bismuth oxalate in absence of air.

$$(BiO)_2\ C_2O_4 \rightarrow 2BiO + 2CO_2$$

When treated with excess of stannous chloride it is reduced to metallic bismuth.

Bismuth (III) Nitrate Oxide (Bismuthyl nitrate $BiONO_3$): A white insoluble solid obtained by hydrolysing bismuthyl chloride with water. It is used in medicine for treatment of diarrhoea and dysentery.

Bismuth oxy Chloride or Bismuth oxide Chloride (BiOCl): A white crystalline insoluble powder. It is us in manufacture of artificial pearls and pigments.

Bismuth Pentaoxide (Bi_2O_5): A brown powder obtained, when chloride is passed for a long time through a suspension of bismuth trioxide in KOH solution, a red coloured potassium bismuthate is precipitated. It is evaporated with dilute nitric acid.

$$KBiO_3 + HNO_3 \rightarrow KNO_3 + HBiO_3$$

$$2HBiO_3 \rightarrow H_2O + BiO_5$$

it is a powerful oxidizing agent and insoluble in alkalis,

Bismuth Sulphate ($Bi(SO_4)_3$): A colourless hygroscopic crystalline substance obtained by dissolving bismuth in hot concentrated sulphuric acid and evaporating the solution to crystallization.

Bismuth Tetroxide or Bismuth Dioxide (Bi_2O_4): A brown powder obtained by passing chlorine through a suspension of the trioxide in an alkali when the trioxide is oxidized to tetraoxide.

$$Bi_2O_4 + 8HCl \rightarrow 2BiCl_3 + 4H_2O + Cl_2$$

It is probably a mixture of bismuth trioxide and pentaoxide.

Bismuth Tribromide ($BiBr_3$): A colourless crystalline substance (m.p. 506 K). Obtained by heating bismuth in a current of chlorine or mercuric chloride.

$$Bi + Cl_2 \rightarrow 2BiCl_3$$

$$2Bi + 3HgCl_2 \rightarrow 2BiCl_3 + 3Hg$$

It is a covalent compound and solubility in organic solvents.

Bismuth Trifluoride (BiF_3): Obtained by heating bismuth trioxide, B_2O_3, in hydrofluoric acid.

Bismuth Tri-iodide (BiI_3): A greyish black substance obtained by the interaction of a soluble bismuth salt in solution. It is hydrolyzed to oxy-iodide, BiOI, on boiling with water.

Bismuth Trisulphide (Bi_2S_3): A black precipitate obtained by passing H_2S through a solution of bismuth salt. It is insoluble in water, alkalis and ammonium sulphide.

Bismuth Trioxide (Bi_2O_3): A yellowish white fusible powder obtained by strongly heating bismuth, its hydroxide or carbonate or bismuthyl carbonate in air

$$4Bi + 3O_2 \rightarrow 3Bi_2O_3$$

it is only slightly soluble in concentrated solution of alkalis owing to the formation of bismuthites.

It is used in the manufacture of optical glass and for glazing porcelain crystalline Bi_2O_3 has a cubic structure.

Bismuth Trichloride ($BiCl_3$): A colourless crystal that forms bismuth oxychloride with water due to hydrolysis. Bismuth oxychloride is used as a white pigment under the name of pearl white.

Bittern: The liquid which is left after sodium chloride has been crystallized out from sea water.

Bitumen: A dark black hydrocarbon manufactured by heating coal. It is used for coating the surfaces of roads.

Bituminous: A form of coal which contains 78% carbon. It is hard like a stone and burns with smoky flame.

Biuret or Carbamoylurea (NH_2 $CONHCONH_2.H_2O$): An insoluble crystalline substance obtained from urea.

Bivalent: Having valancy two.

Bland: Any chemical which does not irritate nor causes discomfort to people, e.g., olive oil.

Blast Furnace: A furnace used for the smelting of iron. A mixture of coke lime stone and crushed ore (Fe_2O_3) is added from the top of the furnace. Heated air sometimes enriched oxygen is forced into the furnace at the bottom. The air coming into the furnace reacts with carbon to form CO; this liberates a lot of heat and allows the furnace to reach a temperature of about 1500°C at the bottom.

Blasting Gelatin: A jellylike mixture of guncotton with nitroglycerin. A very powerful explosive.

Bleaching Powder ($CaOCl_2$): A white solid that can be obtained, by passing chlorine over slaked lime. It gives, chlorine gas with a dilute acid.

Blende: A naturally occurring sulphide ore.

Blowpipe: A piece of apparatus like a pipe ending in a jet.

Body-Centred Cubic Crystal (b.c.c.): A crystal structure in which the unit cell has an atom, ion, or molecule at each corner of a cube and at the centre of the cube.

Bog Iron Ore: An impure form of hydrated, ferric oxide (Fe_2O_3 x H_2O) found in bogs and marshes.

Bohr's Sommerfeld Atomic Model: Bohr's theory could explain the spectra of hydrogen and hydrogen like atoms (i.e., H, H^+, H^{2+}, B^{3+} etc.) but failed to explain the spectra of higher elements. In order to explain this sommerfeld, extended Bohr's work by introducing the concept of elliptical orbits. He assumed that the electrons move in elliptical orbit the nucleus atom being situated at one of the foci of these orbits and circular orbits being simply a special case of elliptical orbits.

While retaining the first Circular orbit suggested by Bohr as such he added one elliptical orbit to Bohr's second orbit, two elliptical orbits to Bohr's third orbit and so on.

Bohr's Theory: Bohr suggested that the negatively charged electrons orbit around the positively charged nucleus in shells of different discrete energy levels. He stated that as long as the atom was in its lowest energy state it would not emit any radiations.

Boiling Point: The temperature at which a liquid begins to boil under normal atmospheric pressure. Boiling at this temperature starts because, the vapour pressure above die surface of tire liquid equals that atmospheric pressure. At sea level water boils at 100°C. At higher altitudes the pressure is lower and hence water boils at lower temperatures making cooking difficult.

Boltzmann Constant (K): A fundamental constant. It is the gas constant per molecule, equal to R divided by Avogadro's number

$$k = 1.381 \times 10^{16} \text{ ergs degree}^{-1} \text{ molecule}^{-1}$$

Bomb Calorimeter: A calorimeter which is employed to find out heat of reaction of constant volume. A strong constant volume vessel capable of withstanding high pressure and mainly used for finding out heat of combustion.

Bond: The force holding atoms, or group of atoms together.

Bond Energy: The energy needed for the dissociation of a particular bond in a specified substance.

Bond Polarity: A measure of the difference in ability of the two atoms in a chemical bond to attract electrons.

Borax Bead Test: A chemical test to detect the presence of some metals. A bead of borax fused in a wire loop will react chemically with the salts of number of metals, often producing colours which help to detect the metal, e.g., manganese compounds give a violet bead, cobalt a deep blue.

Bordeaux Mixture: A mixture of cupric sulphate, $CuSO_4$, calcium oxide, CaO and water. It is used as fungicide.

Boric Acid (H_3BO_3): White crystalline, lustrous flake, having a soft soapy touch. Its water solution is a weak acid. It is used in the manufacture of enamels, glazes for pottery and its solution is used as an eyewash under the name "boric lotion". It is a mild antiseptic.

Borneol or Bornyl Alcohol: A white optically active translucent solid, m.p. 210.5°C. It is used in the preparation of synthetic camphor and in perfumes.

Born-Haber Cycle: A cycle showing how a reaction or a physical process may be considered to be made up of several component steps.

Bornite (Cu_5FeS_4): An important ore of Cu. It is dark brown in colour. Generally occurs with chalcopyrite $CuFeS_2$.

Bornyl Acetate ($C_{10}H_{17}CO\ OCH_3$): A colourless liquid, b.p. 223°C, with a camphor like odour. It is used in the manufacture of perfumes and as a plasticizer.

Boron: A nonmetallic chemical element belonging to group 13. It exists in two allotropic forms, crystalline and amorphous. The parent boron is prepared by the thermal decomposition of $BI_3(g)$ on a heated tantalum filament at about 900°C. It is inert at room temperature to all except powerful oxidising agents. It is used a deoxidizer for casting copper and for making boron steels which are extremely hard and are widely used as control rods in atomic reactors.

Symbol : B; a.n. 5; m.pt. 2300°C; b.pt. 2550°C.

Boron Carbide (B_4C): A very hard black crystalline compound, m.p. 2450°C. It is used in nuclear reactors as moderator and as an abrasive.

Boron Hydrides: The boron hydrides are called boranes with general formula BnHn + 4 and Bn Hn + 6. The well known boron hydrides are : diborane (B_2H_6), tetraborane (B_4H_{10}), pentaborane (B_5H_9), hexaborane (B_6H_{10}), nonaborane (B_9H_{15}) and decaborane ($B_{10}H_{16}$).

Boron Nitride (BN): A slippery white solid obtained by the action of nitrogen on boron. It exists in two structural different forms; hexagonal and cubic form. It is used as refractory material, in thermal insulation of furnaces, as lubricant and electrical insulator for high voltages.

Borosilicate: A boron-containing silicate found naturally, e.g., $CaB_2Si_2O_8$ and obtained as glasses by fusing together $B_2O3.SiO_2$ and a metal oxide. Pyrex glasses are borosilicate glasses,

Boron Trichroride (BCl_3): A colourless mobile liquid, b.p. about 180°C and m.p.–107°C. It fumes in moist air. It is used in the extraction of boron,

Boron Trifluoride (BF_3): A colourless gas in dry atmosphere that gives white dense fumes in moist air, It is used for esterification of alcohols, ether for acids and for polymerization of alkenes. It is a Lewis acid (electron acceptor).

Bosch Process: An industrial method to manufacture hydrogen. Water gas, a mixture of carbonmonoxide and hydrogen, is mixed with steam and passed over heated catalyst. The carbon monoxide is converted to carbon dioxide and hydrogen with steam. The CO_2 is then removed by dissolving it in water under pressure.

Boson: A class of elementary particles behaving according to the Bose Einstein statistics. Examples of bosons are electrons and muons. All, bosons have an integral spin.

Boyle's Law: At constant temperature, the product of the volume and pressure of a given amount of gas is a constant

$p \times v$ = constant (at constant temperature)

Bragg's Equation: This equation, was derived, by W.L. Bragg gives essential relationship that must exist between n, l, q and d in order for diffracted X-ray beam to have maximum intensity

$$n\lambda = 2d \sin \theta$$

The angle of diffraction for which $n = 1$ is known as principle angle. The Bragg equation when applied to diffraction experiment on a crystal with known interplanar spacings (value of d), may be employed to calculate the wavelength of X-radiation used.

Bragg's Law: Defines the relationship between wave-lengths of X-rays falling on a crystals, sparing of planes of atoms in the crystal and the angle of incidence of X-rays, all of which contribute to X-ray diffraction.

Braneling step: A step in a chain reaction that produces more free radicals than it consumes.

Brass: An alloy containing 67% copper and 33% zinc. It is ductile and used for making hardware items.

Breeder Reactor: A nuclear-fission reactor which produces more fissionable fuel than it consumes.

Bredt's Rule: A double bond cannot be added at the bridge head carbon of a bridged bicyclic or polocyclic system with a small or medium sized rings.

Bremstravahlung: A spectrum of X-rays of different intensities, that is, wavelengths and energies produced when fast moving electrons are suddenly stopped or showed down on striking a metal.

Brine: Concentrated NaCl in water. NaCl is generally extracted as brine.

Brittle: A solid material or substance which breaks into small pieces under a force. e.g., glass.

Bromic (I) Acid (Hypobromous Acid, HBrO): A pale yellow liquid prepared by reacting mercury (II) oxide with bromine. It is a good bleaching and oxidising agent,

Bromide: A compound of an element and bromine, The metal bromides are generally ionic in nature and nonmetals are covalent in nature.

Bromide Paper: A photographic paper having silver bromide, AgBr.

Bromine: A darkred liquid having a powerful chlorine like odour, belongs to group 17. It is fairly soluble in water and very soluble in alcohol, ether and carbon disulphide. It occurs in combined state as bromides of sodium, potassium and manganese in sea water. It is prepared in the laboratory by heating a mixture of concentrated sulphuric acid, manganese dioxide and bromide in a retort.

Bromine Acid ($HBrO_3$): Obtained by treating aqueous barium bromate with a calculated quantity of dilute sulphuric acid.

$$Ba(BrO_3)_2 + H_2SO_4 \rightarrow BaSO_4 + 2HBrO_3$$

The aqueous solution of bromic acid decomposes on heating into bromic and oxygen and is therefore, a good oxidizing agent and bleaching agent.

$$2HBrO_3 \rightarrow H2O + Br_2 + 5O$$

Bromine dioxide (BrO_2): A yellowish solid prepared by the action of an electric discharge on Br_2 and O_2 gases at low temperature and pressure. It is less explosive than ClO_2 and on hydrolysis gives bromide and bromate.

$$6BrO_2 + 6NaOH \rightarrow NaBr + 5NaBrO_3 + 3H_2O$$

Bromine Trifluoride (BrF_3): A colourless fuming liquid (b.p. 400 K). It is highly reactive and many metals and nonmetals when dropped in it catch fire.

Bromobenzene or Phenyl Bromide, (C_6H_5Br): A colourless heavy liquid (b.p.428 K). It is obtained by bromination of benzene in the presence of a little iron powder. It closely resembles chlorobenzene.

Bromethyl Blue: An indicator with p^H range 6.0 (yellow) to 7.6 (blue).

Bronze: An alloy of copper and tin. The tin hardens and strengthens the Cu.

Brownian Motion: The random zigzag movement of the fine particles suspended in a fluid, for example, of smoke in air, Robert Brown found that the suspended particles move because they are constantly jostled by moving invisible atoms or molecules in air or fluid.

Brown Ring Test: A test used for the detection of nitrate. A freshly prepared ferrous sulphate is added to the salt solution of nitrate. Then from the side of test tube, we add concentrated sulphuric acid. A brown ring is obtained due to the formation of $[Fe(NO)]SO_4$ (nitroso ferrous sulphate).

Brown oil of vitriol (BOV): A commercial grade of sulphuric acid containing about 77% by volume of H_2SO_4.

Buchner Funnel: A funnel, generally of porcelain, with a flat circular base perforated with a small hole. It is used for filtering by suction.

Buffer: A chemical used to control changes in hydrogen ion concentration during a chemical reaction. It is usually a mixture of weak acids and their salts. They can control pH because they can react with acids and bases.

Bunsen Cell: A type of primary cell in which the positive electrode is formed by carbon plates in a nitric acid solution and the negative electrode consists of zinc plates in sulphuric acid solution.

Buson: A class of elementary particles behaving according to the base. Eimtem, realistic. They are characterised by their pins.

Bunsen Burner: A burner for coalgas used in laboratories. It consists of a metal tube with an adjustable air valve for burning a mixture of gas and air. It is named after R, W, Bunsen and is the first sotters burner in which supply of air is regulated to control the burning flame.

Burette: A long glass tube graduated from 0 to 50 ml, each ml being divided further into ten equal parts, At the bottom. It is provided with a glass tap of pinchcock arrangement. It is used in volumetric analysis to determine the volume of liquids.

Butane (C_4H_{10}): A colourless gas having b.p. 44 5°C. It can be obtained by the electrolysis of sodium acetate and sodium propionate. It is used as a portable supply of fuel, It belongs to homologous series of alkanes.

Butanal or Butyraldehyde ($CH_3(CH_2)_2CHO$]: A colourless inflammable liquid, b p 75.7°C. It is used in rubber industries and in plastics.

Butanedioic acid of Succinic acid ($HOOCCH_2CH_2COOH$): A crystalline solid (m.p. 458K) obtained on a large scale as a by-product during distillation of amber. It is soluble in water and alcohol.

Butanoic Acid (Butyric Acid, C_3H_7COOH): A colourless unpleasant smelling liquid. It is miscible with water, ethanol and ether.

Butanone (Methyl Ethyl Ketone, $CH_3COC_2H_5$): A colourless liquid obtained by the oxidation of butanol-2 with air using silver catalyst at 520 K. It is used as solvent for vinyl resins and synthetic rubber.

Butanol (C_4H_9OH): It exists in two forms, butane -1-ol [$CH_3(CH_2)_2CH_2OH$] and butane-2-ol [$CH_3CH(OH)$ CH_2 CH_3]. Both are colourless volatile liquids used as solvents.

Butyl Group: The straight chain alkyl group $CH_3(CH_2)_2$ CH_2—.

2-Butyne-1,4 diol or Butyne Diol ($HOCH_2{=}CCH_2OH$): A white solid m.p. 58°C obtained by high pressure reaction between ethyne and methanol. It is used in electroplating and in paint and varnish removal.

C

Cadmium: A transition metal obtained as a by-product during the extraction of zinc. It is used to protect other metals from corrosion, as a neutron absorber in nuclear reactors, and in alkali-type batteries.

Symbol : Cd; a.n. 48; m.p. 320°C; b.p. 765°C; r.d. 8.7.

Cadmium Carbonate ($CdCO_3$): A white precipitate obtained by adding any soluble carbonate to the solution of a cadmium salt. It is a white amorphous powder. It changes into CdO and CO_2 on heating. It is used a laboratory reagent to detect the presence of sulphide ion.

Cadmum Chloride ($2Cd\ Cl_2.5H_2O$): Prepared by dissolving the metal, its oxide or carbonate in hydrochloric kid. It is neither deliquescent nor is hydrolysed in aqueous solutions but is efforescent. It is a poor electrolyte.

Cadimum Cyanide ($Cd(CN)_2$): Prepared when potassium cyanide is added to the solution of a cadmium salt, a precipitate of $Cd(CN)_2$, is formed.

Cadmium Hydroxide ($CD\ (OH)_2$): A white precipitate obtained by adding an alkali to the solution of a cadmium salt.

$$Cd^{2+} + 2OH^{-}\ Cd(OH)_2$$

It is insoluble in caustic alkalis but dissolves in ammonia.

Cadmium Oxide (CdO): A dark brown powder obtained by heating me in air or igniting its carbonate or the nitrate. It sublimes at 700°C and is easily reduced by hydrogen or carbon monoxide. It is insoluble in water but soluble in acids forming cadmium salts.

Cadmium Oxy-Acid Salts: Colourless salts obtained by the solution of the oxide for carbonate in the appropriate acid.

Cadmium Sulphide (CdS): A bright, yellow amorphous powder obtained by passing hydrogen sulphide into the solution of cadmium salt. The colour may vary from lemonyellow to orange depending on the conditions of precipitation, It is insoluble in water and dilute acids but dissolves in concentrated acids. It is used as pigment under the name cadmium yellow.

Caesium: A soft silvery white metal belonging to group 1. It is highly reactive, It alloys are used in the manufacture of radiotubes and photoelectric cells. An alloy of caesium and silver is used in the electric eye used in television. It is also used in atomic clocks as a standard of time.

Symbol : Cs; a.n. 55; m.p. 28.5°C; b.p. 690°C; r.d. 1.87.

Caesium Chloride (CsCl): A typical alkali halid obtained from CS_2CO_3 and HCl. Each ion has a coordination number eight.

Caffeine or 1, 3, 7-Trimethy-lxanthine ($C_8H_{10}O_2N_4$): A colourless crystalline silky solid obtained when uric acid is treated with excess of methyl iodide in alkaline solution when it gives trimethyluric acid. This is heated with phosphoryl chloride to get chloro caffeine which on reduction with hydriodic acid yields caffeine. It is also found in plants such as tea, coffee and cola.

Calamine: A zinc mineral originally either zinc carbonate $ZnCO_3$; or zinc silicate, $2ZnO.SiO_2.H_2O$ It is used for a skin preparation, in dermatitis and used either in lotion or as a powder for sunburn.

Calcination: A process of heating an ore to expel water from a hydrate or hydroxide, or carbon dioxide from a carbonates.

Calcium: A silvery white lustrous metal belonging to group 2. It is as hard as tin, and is malleable and ductile. The important minerals of calcium are dolomite, $CaCO_3.MgCO_3$, gypsum, Ca $SO_4.2H_2O$; limestone (chalk or marble), $CaCO_3$. It is obtained by the electrolysis of calcium chloride and calcium fluoride (16%) in an electrolytic cell equipped with two graphite anodes and a hollow iron cathode. It is an active metal next to alkali metals. It combines with oxygen to form oxide and hydrogen to form hydride.

It is used for removing sulphur from petroleum, for making high class steel, in hardening load bearing alloys and for removing bismuth from lead. It is also used to remove traces of water from alcohols.

Symbol : Ca; a.n. 20; m.p. 851°C; b.p. 1487°C; r.d. 1.54.

Calcium Bisulphite (Ca(HSO3)2): Obtained in solution when sulphur dioxide is passed through milk of lime.

$$Ca(OH)_2) + 2SO_2 \rightarrow Ca(HSO_3)_2$$

It is used for removing lignin of the wood pulp and in the brewing industry as an antiseptic.

Calcium Chloride ($CaCl_2$): A white solid which occurs in a number hydrated forms and is readily available as byproduct of the Solvay process, Anhydrous form is used for desiccation of all gases except ammonia and hydrogen sulphide.

Calcium Carbide (CaC_2): A hard crystalline while solid in pure form. The commercial product is generally grey, it is obtained when lime and powdered coke are heated in an electric furnace at about 3000°C.

$$CaO + 3C \rightarrow CaC_2 + CO$$

It is used as a fertilizer under the name of nitroleum.

Calcium Carbonate ($CaCO_2$): In nature exists in three chief forms chalk, limestone and marble. While the last two are hard rocks, the first one is a soft white amorphous mass. Limestone is used for the preparation of lime and cement and as a flux in smelting of ores. Marble is used in the preparation of carbon dioxide and in buildings.

Calcium Cyanamide ($CaCN_2$): A pure form is colourless, but the commercial products is a grey powder. It is formed by heating calcium carbide in the atmosphere of nitrogen.

$$CaC_2 + N_2 \rightarrow CaCN_2 + C$$

It is used as a fertilizer, and in the preparation of urea.

Calcium Cyclamate [C_6H11 $NHSO_3)_2$ Ca2H_2O]: A white crystalline soluble powder, formerly used as sweetening agent in soft drinks.

Calcium Hydrogen Carbonate [Calcium Bicarbonate, Ca $(HCO_3)_2$]: A solid obtained when a carbon dioxide is passed through a suspension of calcium carbonate in water

$$CaCO_3 + H_2O + CO_2 \rightarrow Ca(HCO_3)_2$$

It causes hardness in natural water.

Calcium Hydroxide [Slaked lime, $Ca(OH)_2$]: A white solid that dissolves sparingly in water to give the alkali called lime water. It is obtained by adding water to the oxide, a process known as slaking, which evolves much heat. If just sufficient water is added, so that the oxide turns to fine powder, the product is slaked lime. If more water is added, a thick suspension called milk of lime is formed. Milk of lime is used for white-washing and when mixed with sand and water, it forms mortar which is used as a building material.

Calcium Nitrate [$Ca(NO_3)_2$]: A very hydroscopic salt is obtained when dilute HNO_3 is added to milk of lime or suspension of calcium carbonate in water. It is used as a fertilizer.

Calcium oxalate [$(COO)_2$ CaH_2O]: Obtained as white precipitate adding ammonium oxalate solution to a solution of calcium salt in neutral or alkaline medium. On ignition, it first gives the carbonate and then the oxide.

$$(COO)_2Ca \rightarrow CaCO_3 + CO$$

$$CaCO_3 \rightarrow CAO + CO_2$$

Calcium Oxide (Quick Lime, CaO): A white amorphous powder obtained by heating limestone in vertical rotary kiln to 800-1000°C.

$$CaCO_3 = CaO + CO_2\uparrow$$

It is mainly used for the manufacture of bleaching powder, calcium carbide, sugar, glass, washing soda and caustic soda by causticizing process. Large amount is used for making building material like mortar and cement.

Calcium Peroxide (CaO_2): A white crystalline precipitate obtained when lime water is treated with hydrogen peroxide. It is less stable than barium peroxide. It is used for bleaching and antiseptic purposes.

Calcium Phosphate [$Ca(PO_4)_2$]: A white precipitate obtained by the addition of soluble phosphate to alkaline solution of a calcium salt.

$$3C\ a^{2+} + 2PO^{3-}{}_4 = Ca_3(PO_4)_2\downarrow$$

It commonly occurs in bones and mineral phosphorite. It is used in the manufacture of phosphorus, phosphoric acid and calcium super phosphate.

Calcium Silicates: A variety of compounds including native minerals consisting of calcium oxide and silica in various molecular ratios, e.g., calcium metasilicate, $CaSiO_3$ and calcium orthosilicate, $Ca_2SiO_{4.}$

Calcium Sulphate ($CaSO_4$): It exists in two forms: Anhydrous and dihydrate. It is very sparingly soluble in water. It occurs in nature as anhydrite $CaSO_4$ or gypsum $CaSO_4.2H_2O$.

Calcium Scale: A temperature scale in which the temperature of meeting pure ice is take as 0° and the temperature of boiling water, 100° (both at standard pressure). The oidg celcius (°C) is equal to the taking. This was known as the centigrade scale until 1948, when the present name became official. Celcius original scale was inverted.

Calcium Sulphide (CaS): Obtained from SO_2 and lime water or $CaSO_4$ and charcoal at 1000°C It is used in the manufacture of luminous paints and in cosmetics.

Calcium Sulphite ($CaSO_3$): Obtained from SO_2 and lime water or $CaCl_2$ and a sulphite as precipitate. It is used for dissolving the lignin in wood pulp during paper manufacture for sterilisation.

Calgon: A trade name given to a complex salt, sodium hexametaphosphate $(Na\ PO_3)_6$ or $Na_2(Na_4P_6O_{18})$. It is used in softening hard water for boilders.

Calibration: A check on the accuracy and the prescision of the stated, capacity of a volumetric flask or pipet, or of the stated mass of some weight. Any check on the accuracy and precision of a stated mass of some weight.

Californium: A radioactive transuranic element of the actinoid series metals, not found naturally on earth. Californium 252 isotope is used as an intense source of neutrons.

Symbol : Cf, a.n. 98.

Caloric: Symbol cal. A unit of energy approximately equal to 4.2 joule. It was formerly defined as the energy needed to raise the temperature of one grace of water by one degree celcius.

Calorie (Cell): The caloric is the energy that will raise the temperature of 1.0 g of water from 14.5 to 15.5°C. The calorie is 1000 cal (or I kilo calorie), and is the term used to describe the energy content of foods. I cal = 4.184 J.

Calorific Value: The energy content of a fuel defined as the energy released in burning unit mass of the fuel.

Calorimeter: Instrument for determining the quantity of heat evolved, absorbed or transferred. In its simplest form, a calorimeter consists of an open cylindrical vessel of copper or other substance of known specific heat along with a stirrer and a thermometer.

Calorimetry: The branch of chemistry concerned with the measurement of heat.

Camoquine: Derivative of 4-ammoquinoline. It is less toxic and 3-4 times as active as quinine. It is a highly supressive drug in vivax material and falciparum malaria. Its structure is given below.

Camphor ($C_{10}H_{16}O$): A white crystalline solid with a characteristic smell, m.p. 178°C. It is found in the camphor tree. It is used medicinally as a carminative and stimulant and is a popular remedy for colds and used in the manufacture of celluloid.

Caniine ($C_8H_{17}N$): A colourless oil which boils at 439-440 K. It is found in the seeds and other parts of the spotted hemlock, a shrub growing in wet locations. It is optically active. It has an unpleasant smell and turns brown on exposure to air. It is soluble in water but More soluble in alcohol.

Cannizaro's Reaction: The two molecules of aldehydes are involved in the presence of alcoholic or aqueous alkali in which one molecule is converted into corresponding alcohol, e.g.,

$$HCHO + NaOH \rightarrow HCOONa + CH_3OH$$

Caproic Acid or n-Hexanoic Acid ($C_6H_{12}O_2$): An oil m.p. 3.4°C, b.p. 205°C. It is found as glycerides in the milk of cows and goats, in palm oil and in coconut oil.

Carat: (1) A measure of weight of diamonds and other gems, formerly 3.17 grains (0.2053 g), now standardised as the international carat, 200 Mg. (2) A measure of fineness of gold, expressed as parts of gold in 24 parts of the alloy. Thus, 24 carat gold is pure gold, 18 carat gold contains 18 parts in 24 or has a fineness 750.

Carbazole ($C_{12}H_9N$): occurs with anthracene. It is used in dyestuff preparation.

Carbide: Binary compound of carbon e.g., calcium carbide. CaC_2.

Carbohydrate: A substance in animals or plants consisting of poly hydroxyaldehyde or poly hydroxy ketone or molecules or that can be hydrolyzed to these.

Carbohydrates: A large group of organic compounds composed of carbon, hydrogen and oxygen only with the general formula (C_xH_2O) y and comprising monosaccharides and disaccharides (both sugars) and polysaccharides (starch and cellulose). Carbohydrates play an essential part in the metabolism of all living organism, starch being the principal form in which energy is stored and cellulose being the principal Structural material of plants. Monosaccharides like glucose, fructose and disaccharides like sucrose and lactose are instant source of energy for human body.

Carbolic Acid: (Phenol, or hydroxy Benzene, C_6H_5O H). A colourless crystalline sodium m.p. 43°C, b.p. 182°C which turns pink on exposure to air. It is moderately soluble in cold water but readily dissolves in ether and ethanol.

Carbon: A nonmetallic element belonging to group 14. It is one of the most widely distributed elements. It occurs in combined form, such as, limestone, $CaCO_3$ magnesite, $MgCO_3$, calamine, $ZnCO_3$, spathic, iron, $FeCO_3$: dolomite $CaCO_3$. $MgCO_3$ etc. It is common constituent of all organic compounds. An artificial diamonds is obtained by heating pure sugar charcoal and iron in a graphite crucible to a temperature of about 3000°C in an electric furnace. It burns readily in air to form carbon dioxide. It reacts with conc. sulphuric acid and nitric acid and readily reduces them. There are 3 basic allotropes of carbon : diamond, graphite and the recently discovered buckminster fullerness or bucky balls.

Naturally occurring carbon has isotopes : ^{12}C, ^{13}C and ^{14}C.

Symbol C; a.n. 6; m.p. 3570°C; b.p. 4827°C; r.n. 3.51.

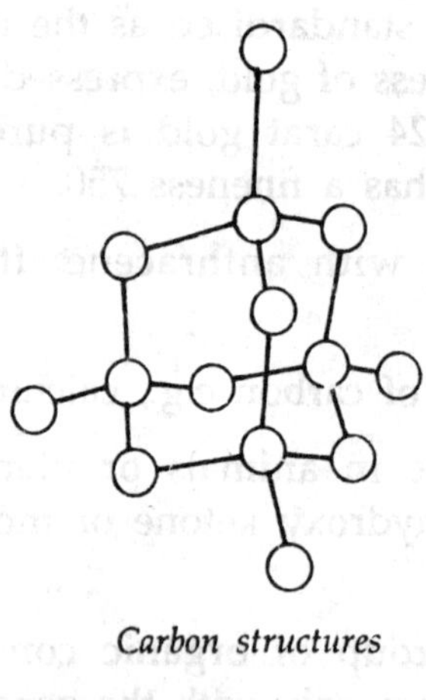

Carbon structures

Carbonate: A salt of carbonic (containing the ion $CO_3^{\ 2-}$ e.g., Na_2CO_3, K_2CO_3, Mg CO_3, $CaCO_3$ etc.)

Carbonion: An intermediate in an organic reaction in which one carbon on atom is electron rich and carries a negative charge. Carbonion are usually formed by abstracting a hydrogen ion from a C—H bond using a base, *e.g.*, from ethanol to form CH_2CHO, or iron organometallic compounds in which the carbon atom is bonded to an clectropositive metal.

Carbohydrates: A class of compound occurring widely in nature and having the general formula type $C_x(H_2O)_y$ (Note that although the name suggests a hydrate of carbon these compounds are in no way hydrates and have no similarities to classes of hydrates). Carbohydrates are generally divided into main classes: sugars and polysaccharides.

Carbon Dating: A process by which the approximate age of an organic material can be determined by measuring the proportion of ^{12}C to ^{14}C; mainly used for dating fossils and archeological finds.

Carbon Dioxide (CO_2): A colourless, carbon dioxide gas obtained by the action of dilute acids on metallic carbonates. It is moderately soluble in water and gives carbonic acid (H_2CO_3). It is used in soda water and aerated water. It is used in the manufacture of sugar, white lead, washing soda (Solvay's process). Solid carbon dioxide known as dry ice is used as coolant. It is also used to extinguish fire.

Carbon Dioxide Cycle: Large amount of carbon dioxide is being formed as a result of combustion and formation of plants and plant products. Animals also breathe in oxygen present in air and exhale carbon dioxide (respiration). Decay of animals after death gives off carbon dioxide. In spite of this, percentages of carbon dioxide and oxygen practically constant in air. This is due to a never ending carbon dioxide cycle in nature. During day time, plants convert carbon dioxide from air to oxygen (photosynthesis).

Carbon Disulphide (CS_2): A colourless flammable liquid prepared form methane and sulphur. It is used as solvent.

Carbonic Anhydrase: An enzyme which catalyzes the reaction between carbon dioxide and water.

Carbon Monoxide (CO): A poisonous, colourless and tasteless gas. It is produced when carbon is burnt with insufficient amount of oxygen,

$$C(S) + \frac{1}{2}O_2(g) \rightarrow CO(g) \quad \Delta H^\circ = -110{,}5 \text{ kJ/mol}$$

It is found in automobile exhausts and mines.

Carbon Tetrachloride (Tetra Chloro Methane, CCl4): A colourless liquid, b.p. 77°C which has a sickly smell. It is insoluble in water but readily soluble in ethanol and ether.

Carbonyl group: The bivalent: C=O group present in aldehydes, ketons and carboxylates.

Carbonyls: Compounds of metals with carbon monoxide e.g., nickel, carbynyl, $Ni(CO)_4$.

Carboxyl Group: The organic group —COOH present in carboxylicacid.

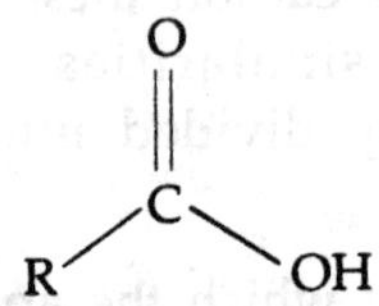

Carboxylic acid

Carbylamine reaction: A qualitative test for primary amines comprising of warming the nitrogen containing compound with trichloromethane (chloroform) in an alcoholic solution of potassium hydroxide; while the resulting bad smell of iscocyanide (RNC) is an indication of primary amino group ($-NH_2$).

$$(CHCl_3 + 3KOH + RNH_2 \rightarrow RNC + 3KCl + 2H_2O)$$

Carboxy Hemoglobin: Carbon monoxide binds very strongly to the iron is hemoglobin to form a complex molecule.

Carcinogen: An agent that cause cancer in animals such as tobacco smoke or ionizing radiation.

Carnallite: A mineral chloride of potassium and magnesium, $KCl.MgCl_2.6H_2O$.

Carnosine ($C_9H_{14}N_4O_2$): An optically active crystalline dipeptide m.p. 260°C; found in muscle tissue.

Carnot's Cycle: An ideal cycle consisting of four reversible processes occurring in succession in a perfect heat engine. These are: (1) an isothermal expansion, (2) adiabatic expansion, and (3) an isothermal compression to such a point that, (4) a further adiabatic compression will return the substance to its original condition.

Cargo's Acid or Peroxo Sulphuric Acid or Permono Sulphuric Acid (H_2SO_5): A white crystalline (m.p. 318 K) solid obtained by reacting one mole of hydrogen peroxide with one mole of chlorosulphonic acid.

$$H_2O_2 + ClSO_3H \rightarrow H_2SO_5 + HCl$$

It is a powerful oxidizing agent. .

Caro's acid: Permonosulphuric acid, H_2SO_5.

Carotenoids: A group of tetraterpenoids. (C_{40}, natural polyenes) that constitute the yellow or red pigments which are widely distributed in plants and animals some of them are hydrocarbons (*e.g.*, carotenes, $C_{40}H_{56}$) others are oxygenated derivatives of carotenes, (*e.g.*, xanthophylls). Sonic of these natural polyenes and their derivatives containing fewer than 40 carbon atoms, but structurally related to carotenoids, are called apocarotenoids, *e.g.*, bixin, $C_{25}H_{30}O_4$ and crocetin, $C_{20}H_{24}O_4$ (at. lipochiontes, chromolipids)

Cast Iron: An alloy of iron which contains 2.5% carbon and other impurities. It is manufactured by smelting iron ores in blast furnace with coke. A very hard and brittle solid, it is used in making stoves, pipes, radiators and toys. It is also used in making gutter pipes and lamp posts.

Castner-Kellner Cell: An electrolytic cell for the manufacture of sodium hydroxide.

Castor Oil: A vegetable oil taken out from the seed of the castor plant. It is used in paint and varnish industry as well as medically as laxative.

Catalysis: A phenomenon in which addition of a small quantity of a substance enhances the rate of reaction manifold, without the added substance called catalyst, itself being consumed.

Catalyst: Substance which alters the rate of a chemical reaction but itself remains unchange. Catalysts are widely used in chemical industry, metals in finely divided states (for hydrogenation), oxides of metals (for dehydration and oxidation), Lewis acids (for alkalytion, acylation and polymerization) are frequently used.

Catenation: The tendency of an atom to form a I *e.g.*, carbon atom.

Cathode: An electrode which pulls the electrons from the external circuit or at which reduction occurs.

Catholyte: The electrolyte near the cathode during clectrolyses.

Cation: A positive ion obtained by removal of electrons from atoms or molecules.

Cationic exchange resins: Polymeric organic compounds containing sulphonic acid ($-SO_3H$), carboxylic acid functions, from which acidic hydrogen can be exchanged with a metal ion (sodium or potassium ions), These sodium or potassium salts of the resin are used as cation exchange resins in ion exchange chromatography and purification of water and other compounds. Sodium alkyl sulphates (Na^+—OSO_2O $CH_2(CH_2)n$ CH_3), sodium alkanes sulphonate ($CH_3(CH_2)_nCH_2SO_3$—Na^+) and sodium alkybenzene sulphonate ($CH_3(CH_2)n(CH_3)CH$ C_6H_4 SO_3^- Na^+ cation exchange resins.

Cell Potential (E cell): The emf that can be produced galvanic cell when no current is drawn from the cell.

Cell Reaction: The overall chemical change d cell or a galvanic cell.

Celsius Scale: Temperature scale on which water freezes at 0°C and boils at 100°C at the presence of 1atm.

Cellulose: A polysaccharide ($C_6H_{10}O_5$). is main constituent of plant cells, It contains parallel unbranched chains of glucose units condensed together into strong fibrous structure. Obtained from wood pulp, cotton and other green plants, it is used for making paper, rayon, plastics, explosives and is an adsorbent for chromatography.

Cellophane: A transparent sheet of cellulose prepared by extracting cellulose xanthate solution into a bath of acid. It used as a wrapping material for food stuffs and cigarettes.

Ceresin: A hard brittle paraffin wax with 70°C-100°C. It is use a melting point in the range as a substitute of beeswax in paints and polishes.

Cesium: A silver white element belonging to group 3. It is found in nature with other lanthanoids, in may mineral including monazite and bastnasite, it is used in alloys, as catalyst and in the glass industry.

Symbol : Cc; a.n. 58; m.p. 799°C; b.p 3430°C; r.d. 6.7.

Cesium Dioxide or Ceria (CeO_2): A white crystalline powder, m.p. 2500°C. It is used in glass polishing.

Chain: When two or more atoms from bonds with each other in a molecule, of atoms results. This chain may be, a straight chain, in which each atom is added to the end of the chain, or may be a branded chain, in which the main chain of atoms has one or more smaller side chains branching off it.

Chain Reaction: A chemical reaction in which many molecules undergo chemical reaction after one molecule becomes activated.

Chalcogens: The elements of group 16 of the periodic table : Oxygen, sulphur, selenium, tellurium and polonium.

Chalk: A naturally occurring form of calcium carbonate ($CaCO_3$) formed originally by marine organisms.

Charcoal: An amorphous form of carbon obtained by heating wood or cellulose or other organic materials in the absence of air.

They are good adsorbents and are used in removing colouring matter from syrups of sugar, brown sugar, glycerine etc.

Charle's Law: The volume of a given mass of a gas is proportional to absolute temperature at constant pressure.

It may be expressed by the equation

$$\begin{matrix} V_1 & V_2 \\ T_1 & T_2 \end{matrix}$$

where V_1 represents the original volume of a given mass of gas at a temperature T_1 and V_2 represents the new volume of the gas at a new temperature T_2.

Chapasite: A nature zeolite, calcium aluminium silicate.

Chelation: The formation of a closed ring of atoms by the attachment of compounds or radicals to a central polyvalent metal ion, e.g., two molecules ($NH_2CH_2.CH_2\,NH_2$) form a chelate ring with a cupric ion as shown below:

Chemical Bond: The force which keeps the atoms in any molecule together.

Chemical Energy: The part of the energy stored within an atom or molecule that can be released by a chemical reaction.

Chemical Equation: A symbolic representation of a chemical reaction in terms of symbols and formulae of various substances involved in the reaction.

Chemical Property: A property which describes the way in which a substance acts when heated, electrolysed or added to other substances so that a chemical change takes place.

Chemical Warfare: A way of fighting wars in which poisonous chemicals, flames smokes and defoliating agents are used, e.g., mustard gas, nerve gas.

Chemistry: The study of the composition of substance and the way their properties are related to their composition.

Chemotherapy: The use of chemicals (drugs) to selectively destroy infections microorganisms without simultaneously destroying the host.

Chemosynthesis: A process by which some organisms (*e.g.,* bacteria) obtain their energy for CO_2 assimilation by the chemical oxidation of simple inorganic compounds.

Chirality: The property of existing in left hand and right hand form.

Chloral Hydrate [$CCl_3CH\ (OH)_2$]: A white crystalline solid, M.P. 57°C obtained from chloral by the action of water. It is used in medicine and in the manufacture of D.D.T.

Chloramine (NH_2Cl): Obtained by the interaction of hypochlorites and ammonia in aqueous solution.

$$NH_3 + KClP \rightarrow NH_2Cl + KOH$$

It is highly explosive liquid. It is a powerful oxidizing agent and destroys the bacteria present in water.

Chloric (VII) Acid (Per Chloric Acid, $HClO_4$): A yellow liquid obtained by distilling in Vacuo potassium Perchlorate with concentrated sulphuric acid.

Chlorides: Binary compounds with chlorine; salts of hydrochloric acid, HCl.

Chlorine: A greenishyellow gas with a characteristic pungent smell belonging to group 17. It is fairly soluble in water and is highly poisonous. It is obtained as a byproduct in the electrolytic process for the manufacture of sodium hydroxide and metallic sodium from sodium chloride. It is highly reactive but less so than fluorine.

It is used for bleaching cotton, linen, wood pulp and paper either in elemental state or as bleaching powder. It is also used in plastics, such as, polyvinyl chloride (P.VC.), extraction of gold, manufacture of organic compounds like ethyl chloride, chloroform, carbon tetrachloride and used in the manufacture of drugs, antiseptics, insecticide, anaesthetics and refrigerants. Also used in purification of drinking water.

Symbol : Cl; a.n. 17; m.p. 130°C; b.p. 34.6°C; r.d. 3.214 kg m^{-3}.

Chlorine Dioxide (ClO_2): A dark yellow pungent smelling gas obtained by passing dry chlorine over silver chlorate heated to 363 K, The issuing mixture of the gases is passed through a freezing mixture when chlorine dioxide condense white chlorine passed on the uncondenced.

$$2AgClO_3 + Cl_2 \rightarrow 2AgCl + 2ClO_2 + O_2 \; 363K$$

It is fairly stable in dark but decomposes in light.

Chlorine Heptaoxide (Cl_2O_7): A colourless oily liquid obtained by dehydration of perchloric acid $HClO_4$ with phosphorous pent aoxide at 263K. It explodes on heating or on percussion. It slowly reacts with water forming Perchloric acid.

$$Cl_2O_7 + H_2O \rightarrow 2HClO_4$$

Chlorine Hexoxide (Cl_2O_3): A red explosive liquid obtained by mixing a stream of oxygen mixed with chlorine dioxide with another carrying ozonized oxygen.

$$2ClO_2 + 2O_3 \rightarrow Cl_2O_6 + 2O_2$$

It is a strong oxidizing agent and explodes in contact with grease.

Chloroacetone (CH_3COCH_2O): A colourless poisonous liquid, b.p., 119°C. It is used in the manufacture of insecticides.

Chlorobenzene (Mono Chlorobenzene, (C_6H_5Cl): A colourless liquid, b.p. 132°C. It is prepared commercially by passing a mixture of benzene vapour, air and hydrogen chloride over heated copper chloride (catalyst). It is used in the manufacture of aniline, phenol and D.D.T.

Chloro Ethane (Ethyl Chloride, C_2H_5Cl): A gas obtained by adding hydrochloric acid to ethylene in the presence of $AlCl_3$. It is used in the preparation of tetraethyl lead, sulphonal and as a refrigerating agent.

Chlorofluro Carbon (CFC): Any of compounds formed from hydrocarbons by replacing some hydrogen atoms by chlorine and some by fluorine. They are mostly unreaction compounds used as refrigerants and aerosol propellants and in making Plastic foams. It is thought that CFCs in the atmosphere may break down to chlorine and attack the ozone layer. For this reason many countries have bounded or restricted their use.

Chloro Ethene (Vinyl Chloride H2C=CHCI): A colourless gas obtained by passing acetylene gas into dilute hydrochloric acid at 340 K in the presence of Hg^{2+}, ions (catalyst).

It is used in the manufacture of plastics.

Chloro Methane (Methyl Chloride, CH_3Cl): A colourless gas obtained by the chlorination of methane. It is used in the manufacture of aniline dyes, as a refrigerating agent, as fire extinguisher and as local anaesthetic.

Chlorophenol (ClC_6H_4OH): A substituted phenol which exists in three isomeric forms. All forms used in the manufacture of dyes.

Chloroquine: Obtained by heating 4, 7-dichloroquinoline with novaldiamine, in phenol solution.

Chromatography: A technique in which separation of various components is achieved by utilization of difference in the degree to which various substances are adsorbed on to the surface of inert materials.

Chrome alum ($K_2SO_4Cr_2(SO_4)_3$ $24H_2O$): A deep violet octahedral crystals obtained by reduction of an acidified solution of potassium dichromate with sulphur dioxide.

$$K_2Cr_2O_7 + 4H_2SO_4 + K_2SO_4 + Cr_2\ (SO_4)_3 + 4H_2O + 3O$$
$$SO_2 + H_2O + O \rightarrow H_2SO_4X_3$$
$$K_2Cr_2O_7 + H_2SO_4\ 3SO_4 \rightarrow K_2SO_4 + Cr_2(SO_4)_3 + 4H_2O$$

It is obtained in large quantities as a by-product in the manufacture of anthraquinol from anthracene.

Chromium: A bluish white transition metal belonging to group 6. It occurs naturally as chromite ($FeO.Cr_2O_3$). It is very hard, the pure metal being harder than glass. The highest oxidation state is +6, It is very stable in air and attacked by acids less readily than iron. It is used for chromeplating and also in the manufacture of steel.

Symbol : Cr; a.n. 24; m. p 1900°C b.p. 2640°C; r.d. 7.19.

Chromium (II) Oxide (Chromous Oxide, CrO): A black powder prepared by the oxidation of chromium amalgam with dilute nitric acid. At high temperature it is reduced by hydrogen.

Chromium (III) Oxide (Chromic Oxide Cr_2O_3): A green powder which is highly stable. It is prepared by heating ammonium dichromate.

$$(NH_4)_2Cr_2O_7 \rightarrow Cr_2O_3 + N_2 + 4H_2O$$

It is used as pigment and gives green colour to glass.

Chromium (IV) Oxide (Chro mium Dioxide, CrO_2): A black solid formed by heating chromium (III) hydroxide in oxygen at a temperature of 300-350°C. It is very unstable.

Chromium (VI) Oxide (Chromium Triozide, Chromic Anhydride, CrO_3): A real crystalline solid is prepared by the action of concentrated sulphuric acid on potassium dichromate. It is used as an oxidising agent, in volumetric analysis and disinfectant with glacial acetic acid.

Chromous (II) Carbonate or Chromous Carbonate ($CrCO_3$): A bluish yellow precipitate obtained by adding sodium carbonate to a cold solution of chromous chloride. It turns reddish brown on heating.

Chromyl Chloride (CrO_2Cl_2): A dark red vapour obtained by heating, electrovalent chloride with potassium dichromate and concentrated sulphuric acid. It is used as an oxidising agent in organic chemistry.

Cineole ($C_{10}H_{18}O$): A colourless Oily liquid terpene b.p. 176.40°C with an odour of camphor. It is used as perfumes and medicine.

Cinchonine ($C_{19}H_{22}ON$): A white optically active solid (m.p. 537 K). Its structure is similar to that of quinine except that the methoxy group is absent.

Cinnabar: A red mineral form of mercury (II) sulphide. It is the main ore of mercury.

Cis: A geometrical arrangement in other which like groups are adjacent to each other.

Cis-trans isomerism (geometrical isomerism): Forms of isomers associated with compounds containing double bond. The like groups in such compounds may be either on the same side of the plane of the double bond (cis-form or Z-form) or on opposite sides (transform or E-form) e.g., maleic acid and fumaric acid are respectively cis (Z) and Trans (E) forms.

Citral ($C_{10}H_{16}O$): A pale yellow oily liquid (b.p. 497501 K) with a pleasant odour of lemon. It occurs in oil of lemon-glass (70-80%) and oils of lime, lemon and citronella. It is widely used as flavouring agent and making synthetic perfumes.

Citric Acid ($HO_2C.CH_2C(OH)(CO_2H)CH_2CO_2H$): A white crystalline acid important in plant and animal cells. It is present in many fruits.

Clapeyron-Clausius Equation: A thermodynamic equation applying to any two-phase equilibrium for a pure substance. The equation states

$$\frac{dP}{dT} = \frac{\Delta Hv}{T\Delta V}$$

where P is the pressure, T is the absolute temperature, ΔHv the molar latent heat of vapourisation and ΔV corresponding change in the molar volume. If one phase is the vapour, the equation changes to,

$$L_n P = C - \Delta Hv/RT$$

where C is a constant for the particular substance and R is the gas constant.

Clathrates: The compounds of one of the two components has an open structure in the crystalline state containing cavities, holes of channels in which the other component trapped, e.g., b-quinol (host) contains enclosed Ar, Kr, Xe as well as O_2, NO, CHO, HCl.

Clausius-Mosotti Law: The molecular polarization (P) of a substance cl molecular weight M, density d and dielectric constant D is,

$$\frac{(D-1)}{(D+2)} \frac{M}{d}$$

Clays: Naturally occurring administrates which form pastes and gels with water.

Clemmensen Reduction: Aldehydes and ketones may generally be reduced to the corresponding hydrocarbons by heating with amalgamated zinc and hydrochloric acid.

Clinker: A hard mass of material which is not combustible, obtained in furnaces and boiler fires. It is composed of calcium silicate formed by fusion of earthy materials in fuels and ores.

Close-packed Structures: The arrangement in which the spheres of the small size are packed so as to achieve maximum density. In this structure, each sphere is in contact with twelve other spheres, that has a coordination number of twelve.

Coagulation: The association of particles (*e.g.,* in coloids) into clusters.

Coal tar: Thick black oily liquid obtained as a by product of coal gas manufacture. Distillation and purification yields, amongst other valuable products, benzene, C_6H_6 toluene, $C_6H_5\,CH_3$ and anthracene, $C_{14}H_{10}$. Pitch is left as residue which is used for metalling of roads and masking of base of stem of trees against some fungal attacks.

Cobalt: A silverywhite transition metal belonging to group I is hard, malleable ductile and slightly magnetic. The two most important ores of cobalt are Samaltite, $CoAS_2$ and Cobaltite, CoAsS. It is used in electroplating, manufacture of high speed cutting tools, permanent magnets. It is also used to prepare many alloys like satellite (contains cobalt, chromium, tungsten an some carbon) and Festel metal (contains cobalt, iron and chromium).

Cobalt Carbonate (CCO_3): A bright red precipitate obtained by adding a solution of sodium bicarbonate saturated with carbondioxide to a solution of a cobalt salt.

Cobalt (II) Chloride or Cobaltous Chloride ($COCl_2$. $6H_2O$): A ruby red monoclinic crystals of hydrated salts obtained by crystallising a solution of cobaltous oxide in hydrochloric acid. Its property of changing colours or h eating is used in synthetic inks. Anything written with cobalt chloride, solution on white paper in invisible ordinarily but becomes visible on heating. It is also used in chemical hydrometers in which the intensity of the blue shade tells the dryness of the atmosphere.

Cobalt Hydroxide [$CO(OH)_2$]: Obtained as a brownish black precipitate when cobalt salt solution is treated with sodium hypochloride or bleaching powder.

Cobalt (II) Nitrate or Cobaltous Nitrate [$CO(NO_3)_2.6H_2O$]: Obtained by dissolving cobalt carbonate in dilute nitric acid and evaporating the solution.

The salt on heating gives nitrogen dioxide.

$$2C(NO_3)_2 \rightarrow 2CO + 4NO_2 + O_2$$

It is used in the detention of aluminium, zinc and magnesium by Charcoal test. It is also used in the production of cobalt pigments used in ceramics.

Cobaltosic Oxide (CO_3O_4): A black powder obtained by igniting cobaltous nitrate.

$$3CO(NO_3)_2 \rightarrow CO_3O_4 + 6NO_2\uparrow + O_2\uparrow$$

Unlike FC_3O_4 is nonmagnetic, It reacts with Cone. HCl liberating chlorine

$$CO_3O_4 + 8HCl \rightarrow COCl_2 + 4H_2O + Cl_2\uparrow$$

Cobalt (II) Oxide or Cobaltous Oxide (CoO): A dark brown or olive green powder obtained by heating cobaltous hydroxide, nitrate or carbonate to 1270K in the absence of air. It is used for preparing cobalt salts, blue glass, blue enamel and a pigment.

Cobalt (III) Oxide or Cobaltic Oxide (C_2O_3): A black powder obtained by gently igniting cobalt nitrate or carbonate in air. It dissolves in acids to form cobaltic salts.

Codon: An individual unit of hereditary instruction that consists of three amines side chains, side by side, on an mRNA molecule.

Coefficients: Numbers in front of formulas in a chemical equation.

Symbol : Co; a.n. 27; b.p. 2880°C; m.p. 1495°C; r.d. 8.902 at 25°C.

Coenzyme (Cofactor): A substance which required along with the enzyme, if an enzymecatalysed reaction is to occur.

Colchicine ($C_{22}H_{25}NO_6$): A yellow crystalline alkaloid, obtained from the autumn crocus, which interferes with the process of mitosis in such a way that it causes an increase in the number of chromosomes in cell. It is used as in artificial method of obtaining new agricultural and horticultural varieties.

Coligative Properties: The properties which depend upon the numerical concentrations of molecules or molecule like particle of solute in the solution and are independent of the nature of these particles, e.g., lowering of vapour pressure ; elevation of the boiling point, depression in freezing points and osmotic pressure.

Collision Theory: A theory of reaction rates that postulates that the rate of a reaction is proportional to the number of collisions that occur each second, between the reactants.

Coke: Coal that has had its volatile components driven off at high temperature, it is mostly carbon.

Colloids: Particles dispersed in a dispersion medium with a size ranging from 10A to 2000A.

Combustion: A chemical change in which addition of oxygen results in the production of heat and or light.

Common Ion Effect: The suppression of the degree of ionisation of a weak electrolyte by the addition of a common ion, e.g., addition of acetic acid to a salt solution of sodium acetate.

In this case the concentration of acetate ions causes a decrease in the concentration of H′ ions and increases the concentration of unionised acetic acid,

Complex Salt: A compound having a complex ion; either the metal or acid radical of a normal salt, or both can form a complex ion, e.g., tetraamine copper (11) sulphate.

Complex: A type of compound in which molecules or ions form coordinate bonds with a metal atom or ion. The coordinating species have lone pairs of electrons, which they can donate to the metal atom or ion. The formation of such complexes is typical of transition metals.

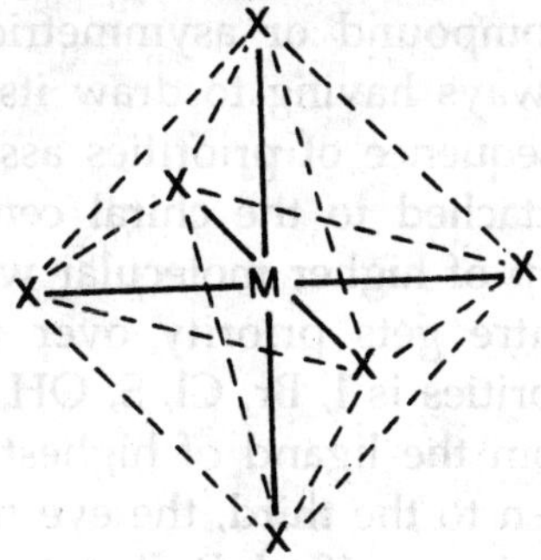

octahedral

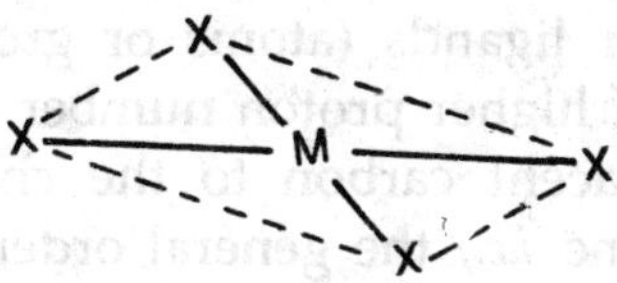

square planar

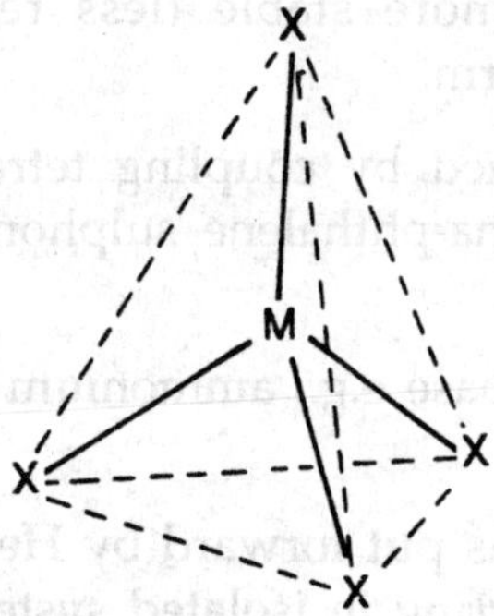

tetrahedral

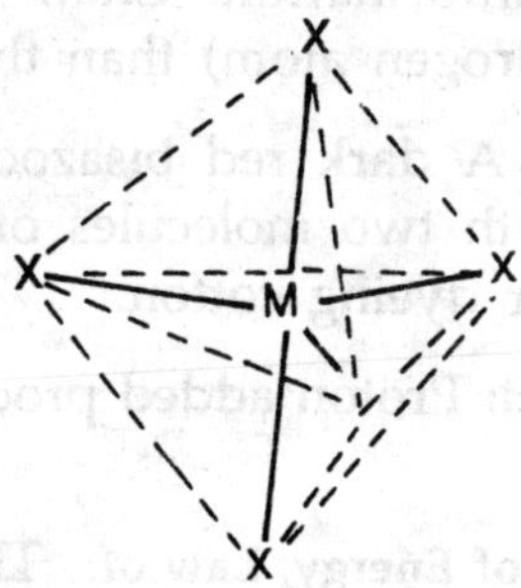

trigonal-bipyramid

Shapes of complexes

Compound: A combination of one or more elements. These substances are homogeneous and their properties and composition are the same throughout their entire mass.

Conductivity: The property by virtue of a substance allows the flow of electric current., The reciprocal of the resistance of a circuit is Called conductivity. The reciprocal of the specific resistance is the specific conductivity or conductance of a substance.

Conductivity of Metals: The electrons in the crystal lattice are free to move the metal making a good conductor.

Configuration: The arrangement of electrons about the nucleus of an atom. Configurations are represented by symbols which contain : (1) An integer, which is the value of the principal quantum number (shell number); (2) A lowercase letter.

Configuration absolute, specification: In order to specify a particular configuration of an optically active compound or asymmetric carbon atom in some simpler way than by always having to draw its picture; prefix R and S are used, following a sequence of priorities assigned to the four ligands (atoms or groups) attached to the chiral centre. The atom of higher proton number or group of higher molecular weight on the adjacent carbon to the chiral centre gets priority over the next lower one *i.e.,* the general order of priorities is I, Br, Cl, S, OH, COOH, CHO, CH_2OH, H. If, in proceeding from the ligand of highest priority to the ligand of second priority and then to the third, the eye travels in clockwise direction, the configuration is specified R (Latin : *rectus,* right); if anticlockwise, the specification is S (Latin *sinister* left)

Conformation: The molecules capable of forming isomers by rotation about a single bond, e.g., cyclohexane exist in two conformations chair and boat conformation. Chair form is more stable (less repulsion between hydrogen atom) than the boat form.

Congo Red: A dark red bisazodye obtained by coupling tetrazotised benzidine with two molecules of 4-aminona-phthalene sulphonic acid. It is used for dyeing cotton.

Conjugate acid: Proton added product of a base e.g., ammonium (NH_3 + H^+)

Conservation of Energy, Law of: The law was put forward by Helmholtz in 1847. It states that in all processes occurring in isolated systems, the energy of the system remains constant.

Constant Composition, Law of: The law was put forward by Proust in 1799. It states that the same compounds always contain the same elements combined together in the same fixed proportion by weight. Carbon dioxide obtained by various methods will have the same ratio by weight between carbon and oxygen, i.e., 3:8.

Conservation of Mass, Law of: The law was put forward by Lavosier in 1774. It states that matter cannot be created or destroyed. When a chemical change occurs, the total mass of the products is the same as the total mass of the reacting substance.

Contact Process: An industrial method to manufacture sulphuric acid. The sulphur dioxide and oxygen is passed over the surface of V_2O_5 or platinised silica, at about 450°C.

$$2SO_2 + O_2 = 2SO_3$$

The sulphur trioxide is cooled and then passed to the absorption tower in which concentrated sulphuric acid is sprayed from the top. Thus, oleum is formed.

$$SO_3 + H_2SO_4 \rightarrow H_2SO_4.SO_3 \text{ ((oleum))}$$

The oleum is finally diluted to get the desired concentration of sulphuric acid.

Continuous Spectrum: A spectrum formed by the light emitted by incandescent solid when it is examined through a prism, *e.g.*, electric light, lime light, arc light etc.

Coordinate Bond (Dative Bond): A special type of covalent linkage in which both the shared electrons are contributed by one atom only and not one by each atom as in a covalent bond.

Coordination Number: The number of units, *i.e.*, donor molecules or ions grouped about a central ion.

Copper: A transition reddish brown metal belonging to group 11. It occurs in nature mainly as sulphide. It is an excellent conductor of heat and electricity, the best conductor next to silver. It is malleable and ductile. It shows poor chemical activity. It is used for making electric cables, electric goods, alloys (brass, bronze) in electroplating, in making ornaments, and salt used in agriculture insecticides and pesticides.

Symbol : Cu; a.n. 29; m.p. 1083°C; b.p. 2582°C; r.d. 8.9.

Copper (I) Chloride (Cuprous Chloride, CU_2Cl_2) A white powder insoluble in water. It is prepared by reducing cupric chloride with zinc dust or sulphur dioxide.

$$2CUCl_2 + Zn \rightarrow CU_2Cl_2\downarrow + ZnCl_2$$

It is used for removing carbon monoxide from other gases and testing acetylene.

Copper (II) Chloride (Cupric Chloride, $CuCl_2$): The crystalline cupric chloride, $CUCl_2$. $2H_2O$ exists in small green crystals, white anhydrous salt is brown in colour. It is fairly soluble in water, The hydrated cupric chloride is used in making pigment and also as catalyst for the preparation of chlorine.

$$4HCl + O_2 \xrightarrow{Cu_2Cl_2} 2H_2O + 2Cl_2\uparrow$$

Copper (II) Hydroxide [$Cu(OH)_2$]: A blue gelatinous precipitate obtained when sodium or potassium hydroxide is added to a solution of copper sulphate.

$$CuSO_4 + 2NaOH \rightarrow Cu(OH)_2\downarrow + Na_2SO_4$$

It is dissolved in ammonia forming deep blue solution known as Schweitzer reagent. This solution can dissolve cellulose which can be reprecipitated on acidification. This reagent is, therefore, used for preparation of artificial silk.

Copper (I) Iodide (Cuprous Iodide, (CuI): A white crystalline powder obtained by adding potassium iodide solution with iodine to copper sulphate solution. It is insoluble in water and not oxidised by air.

Copper Naphthenates Copper (II) salts of carboxylic acid derived from crude petroleum oils and used as wood preservatives.

Copper (I) Oxide (Cuprous Oxide, Cu_2O): A reddish brown powder insoluble in water but dissolves in ammonia solution forming cuproamine ion $[Cu(NH_3)_2]^+$. It occurs in nature as ruby or, ore cuprite. It is used in glass industry and as red pigment for all paints which protection from rusting.

Copper (II) Oxide (Cupric Oxide, CuO): A black hygroscopic powder It is commercially prepared by heating the mineral mulacbite $Cu(OH)_2.CuCO_3$

$$Cu(OH)_2.CUCO_3 \rightarrow 2CuO + H_2O + CO_2$$

It is used for colouring glass light blue.

Copper (11) Nitrate [Cupric Nitrate, Cu(NO3)2]: The hydrated blue crystal $[Cu(N\,O_3)_2.3H_2O$ obtained by dissolving cupric oxide or carbonate in nitric acid. The anhydrous salt (white) can be prepared by treated hydrated crystals with nitrogen pentoxide in nitric acid. It decomposes on heating and is an oxidising agent.

$$2Cu(NO_3)_2 \rightarrow 2CuO + 4NO_2 + O_2$$

Copper (II) Sulphate (Cupric Sulphate or Blue Vitrol, CuSO4.H2O5): A blue crystalline solid obtained by treating sulphuric acid on Copper scrap or turnings placed in a leadlined tower while a current, of air is passed up from below :

$$2Cu + 2H_2SO_4 + O_2 \rightarrow 2CuSO_4 + H_2O$$

It is used in electroplating, in dyeing, in medicine as a fungicide and in the laboratory for the preparation of other copper compounds.

Copper (II) Sulphide (CuS): A black solid that decomposes to copper (I) sulphide on heating.

Corrosion: A reaction in which a metal is attacked by some substance in its environment and converted into unwanted compound, e.g., rusting of iron.

Corundum (every Al_2O_3): A naturally occurring form of aluminitive exide that sometimes contains small amounts of iron and silicon (iv) oxide. Ruby and sappllier are impure crystalline forms, it is used in covarious polishes, abrassives, and grinding wheels.

Coupling: A chemical reaction in two groups or molecules join together. An example is the formation of azo compounds.

Coulomb: The SI unit of electric charge, equal to the charge carried by an electric current of one ampere flowing for one second I C = I As.

Coulometer: An instrument or device of measuring the amount or charge passing in an electric circuit.

Coulometry: Analytical technique involving the measurement of quantities of electricity.

Covalent Atomic Solids: A crystal solid in which the points of crystal lattice are occupied by atoms which share electrons with their neighbours. The covalent bonds are quite strong as a result the solids are quite hard, having a high melting point, and is poor conductor of electricity, e.g., diamond.

Covalent Bond: A bond formed when a pair of electrons is being shared by atoms, e.g., the bond formed by two hydrogen atoms.

Covalent Crystal: A crystal in which lattice positions are occupied by atoms covalently bonded to other atoms at neighbouring lattice sits.

Covalent Radius: The distance between the nuclei of the bonding atoms can be measured, e.g.,

C–C single bond length = 1.54Å

∴Covalent radius of carbon

= = 0.77 Å

Cracking: The thermal decomposition of an organic compound is known as pyrolysis, pyrolysis when applied to paraffins is known as cracking.

Creep: The slow yielding of alloys and metals under a load which does not cause fracture.

Critical point: The conditions of temperature and pressure under which a liquid being heated in a closed vessel becomes indistinguishable from the gas or vapour phase. It temperature below critical temperature the substance can be liquefied by applying pressure but at temperatures above it, this is not possible e.g., for CO_2 critical temperature is 31.1°C and 73.0 atmospheres pressure.

Critical Humidity: The humidity at which the vapour pressure at the surface of a solid or a solution is equal to the partial pressure of water in the atmosphere.

Critical Mass: The minimum amount of fissile material required in a nuclear reactor or a nuclear weapon to sustain a chain reaction.

Critical Pressure: The lowest pressure required to bring about liquefaction of a gas at its critical temperature.

Critical Temperature: The temperature above which a gas cannot be liquified, no matter how great the pressure.

Crum Brown Rule: A guide to substitution in benzene derivatives. This rule says that a substance C_6H_5 a yields the meta disubstituted product if the compound HA can be oxidized directly to HOA; otherwise a mixture of the *o* and *p*-compounds will be formed. It is not applicable universally.

Crystal Field Splitting: The difference in energy between sets of d orbitals in a complex.

Crystal Field Theory: A theory in which it is assumed that the bonds between the central atom and the ligands are purely electrostatic. It helps to explain the magnetic and spectroscopic properties of the complexes.

Crystal Lattice: The repeating symmetrical pattern of atoms, molecule or irons that occurs in a crystal.

Crystalline Solid: A definite rigid geometrical shapes and the constituent molecules, atoms, or irons in them arranged in definite geometrical configuration in threedimensioned network. It has a sharp melting point.

Crystallography: The branch of science which deals with the geometry, properties and structure of crystals and crystalline substances.

Crystal Structure: The particular repeating arrangement of atoms in a crystal.

Crystal System: A division, of crystals based on the shapes of their unit cell, There are mainly seven crystal systems of crystal structure : Cubic, Tetragonal, Rhombic or Orthorhombic, Triclinic, Monoclinic, Trigonal or Rhombohedral and Hexagonal.

Crystal Violet: A bronzy green crystalline solid obtained by heating Michler's ketone with dimethyl aniline in the presence Of $POCl_3$ or $COCl_2$ Wool and silk dyed by it directly.

Cu: Symbol of copper, a metal element having proton number 29 and atomic mass 63.546. It is amicronutrient of plants.

Cubic: Donating a crystal in which the unit cell is a cube. In a simple cubic crystal the particles are arranged at the comers of a cube. The other forms of cubic crystals are facecentred and body-centred.

Cubic: Symbol : G. A unit of radioactivity, equivalent to the amount of a given radioactive substance that produces 3.7 x 1010 disintegrations per second, the number of disintegrations produced by one gram of radium.

Curinal: A highly toxic radioactive silvery, element of the actinoid series of metals. A transuranic element, it is not found naturally on earth but is synthesized from plutonium. Curinal-244 and Curinal-242 have been used in thermometric power generators.

Cupellation: A process in which impure silver is heated in a small bone-ash crucible (cupel) in a blast of air inside a small reverberatory furnace; lead gets slowly oxidised to lithrage, PbO, which is blown away.

Cupric Acetoarsenile or Paris Green $(CH_3COO)_2$ Cu_3.Cu $(AsO_2)_2$ or CU_4 $(CH_3O_2)_2$ (AsO_3): Prepared by boiling verdigris, arsenious oxide and acetic acid together. It possesses a unique light green colour and was used as pigment. Nowadays it is used as an insecticide.

Cupric Arsenite or Scheele's Green ($CuHASO_3$): Obtained as a bright green precipitate by the interaction between ammonium arsenite and copper sulphate solution.

$$(NH_4)_2\ HASO_3 + CUO_4 \rightarrow CuHASO_3\downarrow\ (NH_4)_2SO_4.$$

Used as Green Pigment

Cyanamide Process: An industrial process for fix nitrogen by heating calcium dicarbide in air.

$$CaC_2 + N_2 \rightarrow CaCN_2$$

The product, calcium cyanamide, hydrolyses to ammonia and can be used as a fertilizer. The process is expensive due to the high cost of calcium dicarbide.

Cyanic Acid (HNCO): A mobile volatile liquid. A weak acid, polymerizes to a solid polymer cyanamide above 0°C.

Cyanide: A salt of hydrogen cyanide, containing the cyanide ion (CN).

Cyanohydrin: An addition compound formed between an aldehyde or ketone and hydrogen cyanide. The compound is easily hydrolysed to corresponding hydroxycarboxylic acid, e.g., 2-hydroxypropionitrile formed addition of HCN to acetaldehyde hydrolyses to give 2-hydroxypropionic acid.

Cyanide Process for Gold: The extraction of gold from its ores by dissolving the gold in a solution of potassium cyanide, reducing the resulting potassium aurocyanide, $KAu(CN)_2$ with zinc, filtering off, melting down and cupelling the metal.

Cyanoacetic ester or Ethyl Cyanoacetate ($CN\ CH_2COO\ C_2H_5$): Obtained by treating monochloroacetic acid with potassium cyanide when we get cyanoacetic acid which being an acid, can be esterified. It is used in the synthesis of organic compounds.

Cyanogen or Ethanedinitrile (C_2N_2): A colourless, poisonous gas, very soluble in water. It is combustible and bums with a violet flame to give carbon dioxide and nitrogen.

$$C_2N_2 + 2O_2 \rightarrow 2CO_2 + N_2$$

Cyanogen or Ethanedinitrile (C_2N_2): A colourless, poisonous gas, very soluble in water. It is combustible and bums with a violet flame to give carbon dioxide and nitrogent.

$$C_2N_2 + 2O_2 \rightarrow 2CO_2 + N_2$$

Cyanogen Chloride: A colourless, poisonous gas, very soluble in water, alcohol and ether.

Cyclo Pentane (C_5H_{10}): A colourless liquid, b.p. 49.20°C, prepared from petroleum. It is used as a solvent.

Cyclic Compound: A compound having a ring of atoms. In case the compound contains all the same atoms then it is called homocyclic and if different atoms, it is called hetrocyclic.

Cyclization: A reaction in which a straight chain compound is changed into a cyclic compound.

Cyclobutane or Tetramethylene (C_4H_8): A colourless, pleasant smelling gas (b.p. 258 K). It is more stable than cyclopropane. It is obtained by reduction of cyclobutene with hydrogen in the presence of nickel at 375 K.

Cyclohexane (C_6H_{12}): A colourless liquid obtained by the reduction of benzene with hydrogen in the presence of nickel at 200°C. It is used as a solvent.

Cyclohexanole ($C_6H_{11}OH$): A crystalline soluble substance, m.p. 25.1°C. b.p. 161.1°C. It is used as a solvent.

Cyclopropane (C_3H_6): A colourless gas with a sweetish odour b.p. 34.5°C, obtained by treating 1, 3 dibromopropane with zinc. It is used as anaesthetic.

Cyclopentane or Pentamethlene (C_5H_{10}): A colourless pleasant smelling mobile liquid (b.p. 323K). It is remarkably stable. It occurs in petroleum. On catalytic hydrogenation it undergoes ring fission art gives *n*-pentate.

Cytokinins: A group of natural plant growth regulators containing adenine e.g. zeatin. Synthetic compounds of similar nature have been prepared from adenin and kinetin. These compounds exhibit physiological effects that include bud differentiation of cell cultures and mobilization of organic assimilates within the plant for cell division and cell elongation; retardation of leaf sensecence and induction of protein synthesis.

D

Dalton's Atomic theory: John Dalton a British Physicist and Chemist gave the atomic, theory. The basic Postulates of the theory are that matter consists of small indivisible units called atoms; atoms can neither be created nor destroyed; all atoms of the same element a.e identical and that chemical reactions take place by a rearrangement of atoms.

Dalton's Law of Partial Pressure: In a mixture of gases, the total pressure exerted by the mixture is equal to the sum of the partial pressure exerted by the gases taken separately

$$P_{total} = P_1 + P_2 + P_3 ...$$

Daniell Cell: A galvanic cell having outer copper vessel containing a concentrated solution of copper sulphate. Inside it is a porous pot containing a zinc rod dipping in a dilute solution of zinc sulphate.

d-Block Elements: The groups of elements in which d orbitals are filling up or have just reached a nd_{10} (N + 1)s^2 structure. The elements in this block are called transition elements. There are three series of these in 4th, 5th and 6th period ending with the elements Zn, Cd and Hg respectively.

de-Broglie Relation: The momentum of a particle in motion is inversely proportional to its wavelength, the proportionality constant being Planck's constant. (*h*)

$$\lambda = h/p$$

λ=wavelength.

Debye: A unit of electric dipolemoment equal to 3.33564 x 10^{-30} coloumb meter. It is used in expressing the dipolemoment of molecules.

Debye Huckel Theory: The activity coefficient of an electrolyte depend markedly upon concentration. In dilute solution, due to the coulombic forces of attraction and repulsion of ions tend to surround themselves within atmosphere of oppositely charged ions. Debye and Huckel showed that it was possible to explain the abnormal activity coefficient at least for very dilute solution of electrolytes.

Decahydrate: A crystalline solid containing, ten molecules of water of crystallization per molecule of compound.

Decalin: Obtained when naphthalene is reduced by hydrogen in the present of' nickel as a catalyst first tetralin and then decalin is formed.

It exist in two geometrical forms :

(a) Cis (b.p. 466 K), this is the main product if Pt is the catalyst.

(b) trans (b.p. 458 K), this is the main product if Ni is the catalyst

Decane ($C_{10}H_{22}$): A liquid paraffin hydrocarbon, b.p. 174. 10C, Which is found in various isomeric forms.

Decanoic Acid or Cupric Acid ($C_9H_{19}COOH$): A white crystalline organic acid with unpleasant odour, m.p. 31.5^{0}C. Deconate esters are used in the manufacture of flavouring materials and perfumes.

Decantation: Separation of a solid from a liquid by allowing the former to settle and pouring of the latter.

Decomposition: A chemical reaction in which a compound is split up into compounds with simpler molecules.

Decrepitation: A crystalline solid gives cracking noise on heating due to the removal of water of crystallisation.

Defoliant: A chemical sprayed on plants that causes leaves to fall off prematurely. Agent orange, a defoliant, was extensively used in the Vietnam war to defoliate jungles.

Degenerate: Different quantum states that have the same energy e.g. five d orbitals in a transition metal atom have the same energy but different values of magnetic quantum number m. Differences in energy occur only when a magnetic field is applied to the atom or ligands around the atom which remain no longer symmetrical (i.e. the degeneracy is 'lifted').

Degradation: A chemical reaction in which a molecule decomposes into simpler molecules, e.g., conversion of amide into amine in the presence of bromine and alcoholic KOH.

Dehydration: A process of removing water from a substance, e.g., the dehydration of propanol to propene over hot pumice;
C_3H_7 $CH_3CH{:}CH_2$ + H_2O

Deliquescent: A solid substance which absorbs water from the atmosphere to form a solution.

Delivery Tube: A glass tube joining different pieces of apparatus; it conducts fluid from one piece of apparatus to another.

Delocalisation: The spreading of p electrons over the whole molecule in the compound containing double and single bonds (conjugated system). The abnormal behaviour of conjugated alkadiene is due to delocalisation of electrons.

Delta Iron: An allotropic form of pure iron that exists between 1400°C and the melting point.

Demulsification: The process of breaking the emulsion by adding an electrolyte or a material that would destroy the emulsifier. Heating, freezing or centrifugation may also cause demulsification.

Denaturation: The loss of biological activity in a protein because of disruption of its tertiary structure by heating, by the action of acids or bases or by other means.

Denaturing: The process of, making ethyl alcohol unfit for beverage purposes with the addition of methanol, pyridine or mineral naphtha.

Dendrochronology: Study of the age of tress, and plant materials by measuring the ^{12}C to ^{14}C ratio in the stem rings.

Density (Symbol ρ): The amount of mass in a unit volume of the substance,

Density = mass/volume

the units are generally grams per cubic centimeter (g/cm^3 of $g\ cm^{-3}$).

Deoxyribose Nucleic Acid: Long thread like molecules found in thromosomes and viruses, consisting of two interwound helical chains of polynucleotides. The sugar of all the nucleotides is 2-D-deoxyribose and each nucleotide is characterized by one of the four bases; adenine, cytosine, guanine and thymine. DNA molecules are responsible for storing the genetic code by the order of arrangement of their nitrogenous bases, three bases coding for one amino acid, The structure of a DNA molecule has been found comparable to a twisted rope ladder, the sides of which consist of sugar phosphate chains, the rungs linked nitrogenous bases.

Deploymerization: The breaking down of polymers into their original monomers; the reverse of polymerization.

Depression of Freezing Point: Lowering of the freezing or melting point of a pure compound by impurities. The amount of reduction of freezing point is a colligative property i.e., for solution the lowering of freezing point of the solvent will depend on the modal concentration of the solute and is given by the equation $Dt = K_f C_m$; where K_f (in Kelvin kilogram per mole) is called the freezing point constant or cryoscopic constant, the lowering in freezing point and C_m is the molal concentration of the solute. (Addition of same compound will riot lower the freezing point and that is why the mixed melting point is taken as a criterion for identifying an unknown compound i.e., if a known compound and the compound to be identified have same melting and mixed melting points, then they are identical.

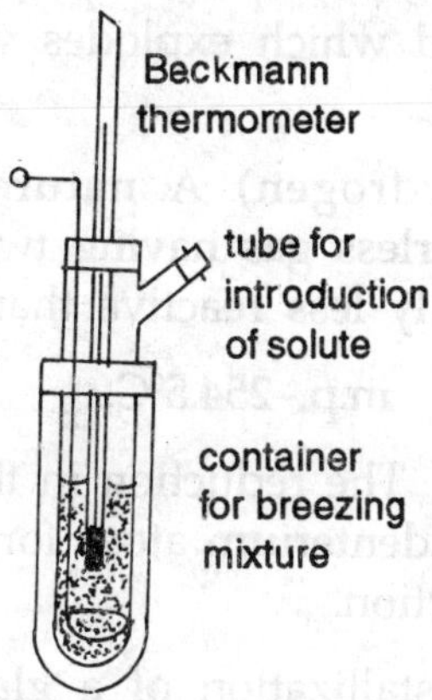

Apparatus for measuring depression of freezing point

Derivative: a compound derived from some other compound, usually keeping the general structure of the parent compound, e.g., phenol is a derivative of benzene.

Desalination: The removal of salts from sea water or brine so as to make it fit for human use.

Desiccant: A material used for drying. Some desiccants reacts chemically with specimens being dried. They are used in a desiccator. In order of effectiveness common desiccants are $CaCl_2$ < CaO < NaOH < MgO < $CaSO_4$ < H_2SO_4 < silica gel < $Mg(ClO_4)2$ < P_2O_5

Desiccator: A glass container used in the laboratory to keep the solids free of moisture. It generally contains calcium oxide or silica gel to absorb moisture.

Destructive Distillation: The process of removing volatile substances form organic compounds in the absence of air, e.g., destructive distillation of wood and coal.

Detergents: A molecule designed along similar lines to soap molecules. They have a long nonpolar carbon chain of 12 to 18 atoms and a highly polar group at one end of the molecule. The advantage that these materials have over ordinary soap is that they do not form insoluble precipitates with calcium, magnesium and iron (iii) ions and result they work equally well in hard and soft water.

Detinning: The recovery of metallic tin from scraptinplate by the action of chlorine that combines with tin to form volatile stannic chloride, $SnCl_2$.

Detonating Gas: A mixture of H_2 and O_2 (molar ratio 2 : I) produced by electrolysis of water and which explodes violently to reform H_2O on ignition.

Deuterium: (Heavy Hydrogen) A naturally occurring isotope of hydrogen. It is a colourless gas having twice the density of ordinary hydrogen. It is chemically less reactive than hydrogen.

Symbol : D; b,p, 249.5^0 m.p.–254.5^0C.

Deuterium Isotope Effect: The reduction in the rate of a reaction caused by the substitution of a denterium atom for a hydrogen at a molecular site undergoing ,he reaction.

Devitrification: The crystallization of a glass (a super cooled liquid) which then becomes opaque.

Dextrin $(C_6H_{10}O_5)^n$: Prepared by the hydrolysis of starch with boiling water under pressure at 520 K. It is a white powder and finds use in sizing paper and in the manufacture of adhesives.

Dextro (d): An optically active compound that rotates the plane of polarised light to right hand side.

Dextrose (Grape Sugar, $C_6H_{12}O_6$): A white crystalline solid (m.p. 419 K), sweet in taste. It is optically active and is dextrorotatory. It is used in medicine and as sweetening agent. It is also used in fruit preservation.

Dextrose (Grape Sugar): Naturally occurring glucose belongs to the sterio chemical series D and is dextrorotatory indicated by the symbol (+). Thus the term dextrose is used to indicate D(+) glucose.

Dewar Flask (Vacuum Flask): A double walled container of thin glass with a space between the walls evacuated and sealed to stop conduction. It is used to store liquid air.

Diagonal Relationship: A relationship in which the first element in a group resembles the second element of the second period and so on.

Dialysis: A process of selective diffusion through a semipermeable membrane. It is used to separate low molecular weight solutes which diffuse through the membrane from colloidal and high molecular weight solutes which do not pass through.

1, 6-Diaminaoxhane [$H_2N(CH_2)_6NH_2$]: An organic compound used as starting material in the manufacture of nylon

Diamond: It is the hardest natural substance (m.p. 7350^{0}C). A carbon atom is bounded convalently to each of the four other carbon atoms situated around it tetrahedrally, it is insoluble in all solvents and a bad conductor of heat and electricity. It is an allotropic form of carbon. It is used in jewellery. Due to its hardness, it is used in cutting glass, as rock drill, photography needles and as abrasive.

Diaphragm Cell : A cell for electrolysis of brine in which the anode and cathode compartments are separated by diaphragm.

Diatomic: A molecule which contains two atoms, e.g., Oxygen (O_2), Chlorine (O_2), Nitrogen (N_2) and Hydrogen (H_2).

Diazoacetic Ester or Ethyl Diazoacetate (CHN_2COO C_2Hs): A yellow oil (b.p. 414 K) obtained by treating a cooled solution of hydrochloride of ethyl glycine with a cold solution of sodium nitrite.

Diazomethane (N_2CH_2): A yellow gas with a peculiar musty odour. It is highly poisonous and highly explosive. It is very useful reagent in organic synthesis.

Diazonium Salt: A compound of the type Ar N_2^+X (the negative ion generally halide or sulphate). Diazonium salts are made by diazotization of aromatic amines and can be isolated by diazotization of aromatic amines and can be isolated at low temperatures. They are important intermediates for preparation of azo dyes, diazoles and triazoles.

Diazotization: The reaction of an aromatic amine (e.g. aniline) with nitrous acid at low temperature (below 5^0 C). $C_6H_5NH_2$ + HNO_2 $C_6H_5N:N^+$ + OH + H_2O The acid is prepared in situ by reaction between sodium nitrite and an inorganic acid, The resulting diazonium ion is susceptible to attack by nucleophiles and provides a method for substitution onto the benzene ring.

Dibasic Acid: An acid which has two active protons, such as sulphuric acid.

Dibenzene Chromium ($C_{12}H_{12}$ Cr): A brown black crystal, m.p. 284-285^{0}C, quickly oxidized by air. It is obtained from $CrCl_3$, C_6H_6, $AlCl_3$ and a reducing agent.

Dibutyl Oxalate ($(C_4H_9OOC)_2$): A colourless liquid, b.p. 243.40C It is used in organic synthesis and as a solvent.

Dichlorine Oxide (Chlorine Monoxide, $C_{12}O$): An orange gas obtained by passing chlorine over yellow mercuric oxide. It is very soluble in water and is very unstable. It is a strong oxidising agent.

Dichloroacetic Acid ($CHC_{12}COOH$): A colourless liquid. On careful hydrolysis with dilute alkali it gives glyoxylic acid. On vigorous hydrolysis with concentrated alkali, glyoxylic acid produced undergoes Cannizaro's reaction to give oxalate and glycollate.

Dichromate (vi): A salt containing the iron Cr_2O_7. Dichromates are strong oxidising agents.

Dicyanodiamide or Cyanoguanidine [$H_2NC(NH)NHCN$]: A white crystalline substance, m.p. 208^{0}C, used in the manufacture of melamine and of barbiturates and other drugs.

Dicyclopentadiene: A colourless solid, m.p. 32°C with a characteristic odour.

Dielectric: A material which is an electrical insulator or in which an electric field can be sustained with a minimum dissipation of power.

Dielectric Constant: The capacity of the solvent to weaken the forces of attraction between the electrical charge immersed in it.

Diels-Alder Reaction: A method of preparing a benzene ring from a diene and a compound containing a single double bond *(e.g.,* maleic acid). Named after Otto Diels and Kurt Alder .

Dienoestrol ($C_{18}H_{18}O_2$): A colourless crystalline powder, m.p. 233°C, obtained by reducing phydroxy propiophenone. It has oestrogenic properties.

Dieterici's Equation: A modification of Vander Waal's equation, in which account is taken of the pressure gradient at the boundary gas. It is written as P(V b) = RTe $^{a/RTv}$ where a and b are constant characteristic of each gas; at relatively low pressures.

Diethyl Malonate or Malonic Ester [$CH_2(COOC_2H_5)_2$]: A colourless liquid (b.p, 472 K) with a Pleasant odour. It is insoluble in water, miscible with organic solvents. It is used to prepare various organic compounds.

Diffusion: The Spontaneous movement and spreading of particles. This could be atoms and solids and occurs from higher concentration to lower concentration. The spread of odours and perfumes in a room occurs due to diffusion.

Differential Titration: Analysis of a sample which contains two or more similarly reacting species which are differentiated, e.g., use of different base or.indicators.

Digitoxin ($C_{41}H_{64}O_{13}$): A white crystalline glucoside, m.p. 252-3°C, obtained from digitals and used as a heart stimulant.

Dihydrate: A crystalline compound which contains two molecules of water of crystallisation per molecule of compound.

Dilatometer An apparatus for measuring small changes of volume of a liquid, solution or solid immersed in a liquid.

Dilute Denoting a solution in which the amount cd solute is two relative to that of the solvent. The term is always relative and includes dilution at trace level as well as the common term bench dilute acid, which means a 2 N solution.

Dilute Solution: Any solution With a small ratio of the quantities of to solvent.

Dimer: A compound or molecule formed by combination or association of two molecules, e.g., dinitrogen tetraoxide (N_2O_4) and aluminium chloride (Al_2Cl_6) form a dimer in vapour.

Dimethyl Amine [(CH3)2NH]: A colourless gas with an ammoniacal smell. It is used in dehairing hides and as a rubber accelerator.

Dimethyl Ether or Methyl Ether [$(CH_3)_2O$]: A gas, very soluble in water. It is used as a refrigerating agent and low temperature solvent.

Dinitrogen (Nitrous Oxide, N_2O Laughing Gas): A colourless gas obtained by heating ammonium nitrate up to 270°C. It is used as an anaesthetic and as a propellane gas in whipped cream bomb.

Dinitrogen Tetra Oxide (N_2O_4): A colourless gas that changes to a pale yellow liquid below 21°C and solidifies below 11°C. It is used as nitrating agent.

Dinitrophenyl Hydrazine: 2, 4-Dinitrophenyl hydrazine, an orange solid commonly used in detection and colorimetric analysis of aldehydes and ketones. Aldehydes and ketones form coloured 2, 4-dinitrophenyl hydrazones with characteristic melting points.

Diol (Dihydric Alcohol, Glycol): An alcohol that has two hydroxyl groups (–OH) per molecule of compound.

Diphenylamine [$(C_6H_5)_2NH$]: A colourless pleasant smelling crystalline solid (m.p. 327 K) obtained by heating phenol with aniline in the present of zinc chloride at 530 K.

Diphosphate (Diphosphine, P_2H_4): A yellow liquid that can be condensed out from phosphine in a freezing mixture. It ignites spontaneously in air.

Dipole: A molecule having partial positive and negative charges separated by a distance.

Dipole Moment: An object or system or molecule that is oppositely charged at two points or poles, e.g., *HF* or magnets.

Direct Dyes: A group of dyes that are mostly azocompounds derived from benzene or benzidene derivatives. They are used to dye cotton, glucose rayon, and other cellulose fibres directly, using a natural bath containing sodium chloride or sodium sulphate as a mordent.

Disaccharide: A molecule which on hydrolysis gives two monosaccharide molecules, e.g., hydrolysis of cane sugar.

Discharge Tube: A vessel generally of glass, which contains two metal electrodes and in which the passage of electricity takes place through a gas or vapour.

Disodium Hydrogen Phosphate (V): (Disodium Hydrogen Orthophosphate, $Na_2H\ PO_4$): A white crystalline solid obtained by neutralizing phosphoric acid with soda ash. It is used as laboratory reagent and a mild purgative in medicine.

Disperse Dyes: Water insoluble dyes, which, when held in time suspension, can be applied to accentuate rayon fabric. The dye, together with a dispersing agent, is warmed to a temperature of 4550°C and the fabric added. By modifying the method of application it is possible to dye Tribel, as well as polyacrylic and polyster fibres. The yellow orange shades are notary amine deterivatives and the green to bluish shades are derivatives of 1-amino anthraquinone certain azo-compounds are disperse dyes and these give a range of colours.

Displacement Reaction: A chemical reaction in which an atom. or group displaces another atom or group from a molecule.

Dissociation: Breakdown of a molecule into two molecules, atoms, radicals, or ions, often the reaction is reversible, as in the ionic dissociation of weak acids in water

$$CH_3COOH + H_2O == CH_3COOH + H_3O^+$$

Distillate: The material generally liquid, that is vapourised and condensed in the process of distillation.

Distillation: A process in which both evaporation and condensation take place side by side.

Dithionic Acid ($H_2S_2O_6$): Obtained by oxidizing sulphurous acid by means of manganese dioxide, ferric hydroxide of cobaltic hydroxide

$$2H_2SO_3 + (O) \rightarrow H_2S_2O_6 + H_2O$$

It exists only in solution and decomposes Oil heating into sulphur dioxide and sulphuric acid.

$$H_2S_2O_6\ 4 \rightarrow H_2SO_4 + SO_2$$

DNA: Deoxyribonucleic acid; nucleic acid that when hydrolyzed, gives deoxyribose, phosphate ion, and chiefly four heterocyclic aminesadenine, thymine, guanine and cytosine; the carrier of the genetic messages of an organism.

Dobereiner's Triad: As early as 1829, Dobereiner classified certain elements in group of three called triads. The three elements in a triad had similar properties and atomic weight; the middle element was mainly an arithmetic mean the atomic weights (if the other two.

Dolomite (Pearl Spar): Natural double carbonate of magnesium and—calcium $MgCO_3.CaCO_3$. A whitish solid that occurs naturally in vast amounts, comprising whole mountain ranges.

Doping: The incorporation of ingerilies within the crystal lattice of a solid so as to alter its physical properties. For instance, silicon, when dropped with boron, becomes semiconducting.

Double Bond: A covalent bond between two atoms that includes two pairs of electrons.

One pair forms a sigma bond and other forms a n, bond.

$$CH_3-\overset{\overset{\displaystyle O}{||}}{C}-CH_3$$

Double Decomposition: A chemical change between two compounds in which two new compounds are formed by mutual exchange of their radicals, e.g., reaction between sodium bromide and silver nitrate

$$NaBr + AgNO_3 \rightarrow AgBr + NaNO_3$$

Double Salt: When equivalent quantifies of certain salts are mixed in aqueous solution and the solution evaporated, a salt may form, e.g., $FeSO_4(NH_4)_2SO_4.(6H_2)$. In aqueous solution the salt behaves as a mixture of the two individuals. These salts are called double salts to distinguish them from complex salts, which yield complex ions in solution.

Dow Process: The process for, the extraction of magnesium from seawater by precipitating $Mg(OM)_2$ by $Ca(OH)_2$ followed by solution of the hydroxide in hydrochloric acid.

Down's Process: A process to manufacture sodium metal and chlorine by the electrolysis of fused sodium chloride, potassium chloride and potassium fluoride mixture in an iron box with graphite anode and iron cathode.

Dry Cell: A cell made of up żinc metal containing a paste of ammonium chloride, MnO_2 as oxidant, The zinc container acts as an anode and carbon rod acts as a cathode. It is covered with a steel case so as to protect it from chemical corrosion.

Dryers: Devices used in chemical processes to remove a liquid from a solid by evaporation, Drying equipment is classified by the method of transferring heat to a set solid, This can be by direct contact between hot gases and the solid (direct dryers), heat transfer by conduction through a retaining metallic wall (indirect dryers), or infrared rays (infrared dryers) group or T-bonds.

Dry Ice: The solid form of carbon dioxide resembling snow flakes. It is used as refrigerant under the name drikoid, in cold drinks and for the manufacture of ice-cream.

Dry Cleaning: The method of cleaning fabrics using solvents, which dissolve dirt at low temperature but do not swell fibres.

Dry Distillation: The heating of a solid to make it give off a vapour; the vapour is condensed to liquid, e.g., distillation of solid calcium ethanoate.

Ductility: The ability of a material to be plastically deformed by elongation, without fracture. A property by which materials can be drawn into wires. Silver and gold are highly ductile.

Duff's Reaction: Salicylaldehyde is obtained by heating phenol with hexam-ethylenetetramine, glycerol and boric acid.

Duma's Method: A method to calculate the percentage of nitrogen, The, nitrogen of an organic compound is estimated by heating organic compound with copper oxide, Then the combustion product is passed over a bright copper spiral. Carbon, hydrogen and sulphur are oxidised to CO_2, water and SO_2 respectively, These can be absorbed in caustic potash solution. From the volume of nitrogen collected over caustic potash, percentage of nitrogen can be calculated.

Duralumin: An alloy of aluminium containing magnesium 0.5%, manganese 0.5%, copper 4% and aluminium 95%.

Dynamic Equilibrium: An equilibrium in which two opposing processes are occurring at equal rates.

Dyne: The former unit of force used in the C.G.S. system. It is equal to 10^{-5} N

Dynel or Vinyon N: A copolymer of acrylonitrile and vinyl chloride containing about equimolar quantities of the two monomers and car be spun into fibre from acetone solution.

Dysprosium: A soft malleable silvery element in lanthanoid series. It is used as neutron absorber in nuclear reactions.

Symbol : Dy; a.n. 66; m.p. 1412°C; b.p. 2562°C; r.d. 8.56.

E

Ecgonine or 3-Hydroxy-2-Tropane Carboxylic Acid ($C_9H_{15}NO_2$): A colourless cry stal'ine material, m.p. 203°C, obtained by the hydrolysis of cocaine with acid.

Edeleanu Process: An extra ction method utilizing liquid sulphur dioxide for the removal of aromatic hydro carbons and polar molecules from petroleum fractions,

Edpoxide: A type of organic compound containing a three membered ring in which two carbon atoms are each bonded to the same oxygen atom.

EDTA (Ethylene Diamine Tetracetic Acid): A compound with the formula $(HOO\ CCH_{22}$ $N.(CH_2)2N.(CH_2C\ OOH)_{2.}$ It is used in getting chalets of transition metals.

Effervescence: The scape of small gas bubbles from a liquid usually as the result of chemical action.

Efflorescence: A process in which a crystalline hydrated solid gives out water of crystallization.

Effusion: The rate at which a gas escapes through an orifice of hole.

Eigen Function: In wave mechanics, the Schrodinger equation may be written us ing the Hamiltonian opera tor H as

$$\mu\psi = E\psi$$

only certain energy values (E) will lead to solution of this equation. The corres ponding values of the wave functions are called Eigen functions.

Einstein: A unit of radiant energy. If every molecule in a gram molecule of a substance absorbs one quantum of light, the total energy absor bed is L *hv*, where L is Avogadro's number, h is Planck's constant and v the frequency of the light.

Einstein's Equation: Emc^2 where E is the energy equivalent of a mass m, and c is the velocity of light. A direct consequence of Eins tein's special theory of *relativity*, this equation is the basis of all nuclear energy.

Elastometer: An elastic sub stance, e.g., a natural or synthetic rubber.

Electric Are Furnace: A steel making furnace in which an electric are provides the source of heat. In direct arc furnaces, the arc is formed between an electrode and the metal being heated, The Heroult furnace is an example of this type, Three graphite or amorphous carbon electrodes

Electrodeposition: The process of depositing a layer of solid (metal) on an electrode by electrolysis. Positive ions in solution gain electrons at the electrode and the deposited as atoms. Copper, for instance, can be deposited on a metal electrode from copper sulphate solution.

Electrolytic: Relating to the behaviour or reactions of ions in solution.

Electrolytic Refining: A method of purifying metals by electrolysis. Copper is purified by making the impure, metal in the anode in an electrolytic with copper sulphate electrolyte, The cathode is a thin strip of pure copper. The copper at the anode dissolves (as Cu^{24} ions) and pure copper is deposited on the cathode. In this particular process, gold and silver are obtained as by products, deposited as on anode sludge on the bottom of the cell.

Electron Pair: Two electrons in one orbital with opposing spin (spin paired), such as the electrons in a covalent bond or love pair.

Electrophile: An electron deficient ion or molecule that takes part in an organic reaction. The electrophile can be either a positive ion (H^+, NO^+_2) or a molecule that can accept an electron pair (SO_3O_3). The 2 electrophile attacks negatively charged areas of molecules, which usually arise from the presence in the molecule of an electronegative atom.

Electric Are Furnace: A steelmaking furnace in which an electric are provides the source of heat. In directarc furnaces, the arc is formed between an electrode and the metal being heated. The Heroult furnace is an example of this type. Three graphite of amorphouscarbon electrodes are used and arcs form between each electrode and the metal charge. In the indirectarc furnaces, heat is produced by a discharge between two electrodes and is radiated into the charge. The Stassano furnace is an example of this type.

Electrical Energy: The energy associated with *electric* charges and their movements. Measured in watt seconds (joules)/or kilowatthours. One kilowatt-hour equals 3.6×10^6 joules or 8.598, 45×10^5 calories

Electrochemical Equivalent: The amount of a substance deposited by I ampere current passing for I second (i.e., I coulomb).

Electro chemical Series: A series telling the activities of metals for reactions that involve ions in solution. In decreasing order of activity the series is K, Na, Ca, Mg, Al, Zn, Fe, Pb, 4, Cu, Hg, Ag, Pt, Au.

Electrochemistry: The branch of chemistry which main', deals with the study of oxidation reduction phenomena.

Electrodes: The terminal through which the current enters or leaves the electrolyte in a cell, A positive electrode is called anode and negative electrode is called cathode.

Electrode Potential: The potential difference the electrode and the solution.

Electroforming: The production, or reproduction, of metal articles by the deposition of a metal upon an electrode during electrolysis.

Electrolysis: A process to separate chemically the constituents of compounds by passage of electric current through electrolytic solutions.

Electrolytes: A liquid or solution of a substance that readily conducts electricity. This group includes acids, bases and soluble salts, such as, HCI, NaOH and NaCl.

Electrolytic Cell: A system consisting of electrode immersed in a solution of electrolytes.

Electrometallurgy: A process of extracting metals from their ores by electrolysis, e.g., extraction of aluminium from bauxite.

Electrometric Titration: Analytic method for following titration by measuring e.m.f. of an inert electrode dipped in the solution.

Electron: An elementary particle which is a constituent of all atoms. It ha a negative charge of 1.60×10^{-19} coulombs and a mass (9.1×10^{-28}g) equal to 1/1838th of that of the hydrogen atom.

Electron Affinity: (Symbol A) The tendency of an atom or molecule accept an electron and form a negative ion. In particular, the energy liberated when one mole of an element is converted to negative ions.

Electron Capture: A nuclear reaction in which a nucleus captures an orbital electron that changes a nuclear proton into a neutron.

Electron Density: The concentration of the electron's charge within a given volume.

Electron Diffraction: A method used to find out the structure of substances, mainly the shapes of molecules in gaseous phase. A beam of electrons directed through a gas at low pressure produces a series of concentric rings on the photographic plates.

Electronegativity: A measure of the ability of an atom in a molecule to attract electrons to itself. There is no simple, direct way to measure electronegativity but Pauling's method is the most frequently used and is based upon experimentally derived values of bond energies.

Electronic Transition: In an atom or molecule, the electrons have certain allowed energies only (orbitals). If an electrons moves from one orbital to another, an electronic transition takes place and there is emission or absorption of energy according to the difference in energy of the two orbitals.

Electron Volt: (Symbol eV) A unit of energy equal to 1.602, 1917 x 10^{-19} joule.

Electrophilic Addition: A reaction involving the addition of a small molecule to an unsaturated organic compound.

Electrophilic Substitution: A reaction in which substitution of an atom Electromagnetic Radiation A form of energy possessing wave character or group of atoms with an electrophile takes place, e.g., nitration of and being propagated through space with a characteristic speed of benzene. 3,00 x 108 m/sec.

Electrophoresis: The migration of electrically charged suspended particles present in a colloidal solution towards the oppositely charged electrode under the influence of electric field.

Electrophilic Addition: A reaction involving addition of a molecule to an unsaturated double or triple bond. The initial attack is by an electrophile on the electron rich system of a molecule e.g. addition of HBr to ethylene :

$$H_2C\ CH_2 + H^+ \rightarrow H_3CCH_2^+ + Br \rightarrow H_3CC\ H_2Br$$

Electrophilic Substitution: A reaction involving substitution of an atom or a group of atoms in an organic molecule with an electrophile as an attacking substituent e.g., nitration of benzene

$$C_6H_{16} + NO_2+ \rightarrow C_6H_5NO_2 + H_+$$

Strong acids or Lewis acids are used as catalysts (cf. nucleophilic substitution).

Electroplating: The process of depositing a thin layer of a metal on another metal, for decoration or protection using an electrolytic process, e.g., the electroplating of steel with chromium.

Electrorefining: The process of purifying metals by electrolysis. The impure metal is made the anode and a plate of pure metal is made the cathode. The electrolyte is the aqueous solution of the metal salt.

Electrovalence Number: The electrovalence number of an element is the charge (including sign) of the ion derived form the element after air electrovalent reaction.

Electrovalency Number: The number of valency electrons that must be lost or gained in order of the element to attain a noblegas configuration.

Element: A substance that cannot be decomposed by chemical means into simple substances. There are 106 elements known out which 92 are naturally occurring and rest of the elements were obtained by bombardment of other nuclei with highenergy particles.

Elevation of Boiling Point: A colligative property of solutions in which the boiling point of a solution is raised relative to that of pure solvent. The elevation is directly proportional to the number of solute molecules introduced rather than the specific aspect of the solute composition.

Elimination Reaction: A reaction involving the, removal of a small molecule, often water or hydrogen chloride, from an organic molecule to given an unsaturated compound, an example is the elimination of a water molecule from an alcohol to produce an alkene. An eliminating reaction is often in competition with a substitution reaction and the predominant product will depend on the reaction condition.

Eluent: The solvent used to separate a mixture in column chromatography.

Elution: The removal of an adsorbate from an adsorbent by dissolving it in a liquid (the eluent), The resulting solution is called the eluate.

Elutriation: The method of separating a material into fractions of various sizes by permitting it to settle against upward moving stream of fluid, generally air or water,

Emetine ($C_{29}H_{40}N_2O_4$): A white powder (m.p. 74°C) obtained by the methylation of caphaeline. It is used against amoebic dysentery.

Empirical Formula: The simple whole number ratio of the atoms in various elements present in a compound, e.g., empirical formula of Hydrogen peroxide is HO.

Emulsifying Agent: A substance, small quantities of which help to form or stabilize an emulsion.

Emulsion: A dispersion of tiny droplets of one liquid in another liquid. There are two types of emulsions : Oil in water and water in oil.

Enamels: A hard material containing over 90 per cent calcium and magnesium salts which provides protection corrosion or decay. Naturally present on teeth, it is applied to pottery, ceramics and metals.

Enantiomer: A compound whose structure is not super impossible to its mirror image.

Enantiotropy: The reversible transformation of one form or one allotrope into another at a definite transition temperature, e.g., HgI_2 = yellow $Hg1_2$ at 126°C.

Endosperm: Nutritive food storing tissue which surrounds the embryo during development. In some seeds the endosperm is entirely absorbed by the embryo when the seed is fully developed e.g. in pea and bean, in others it remains and is absorbed during germination' e.g. and other grasses, onion, castor etc.

Endothermic Reaction: A reaction which proceeds with the absorption of heat, e.g. decomposition of potassium chlorate into potassium chloride, and oxygen gas.

End Point: The point in a titration usually indicated by a change of colour of an indicator, at which a particular *reaction* is completed.

Energy: The capacity of a system to do work. It is easy to divide energy into kinetic energy (energy due to Motion) and potential energy (stored energy). Work and energy have the same unit joule (J).

Energy Levels: An atom, molecule, or *nucleus* can exist only in certain definite stages (quantum states) each having definitive energies. Thus, for each different atom, molecule, or nucleus, there exists a series of energy levels corresponding to these permissible states. The lowest stable energy level is referred to as the ground state; systems at higher energy levels than the ground state are said to be excited.

Enol: An Organic compound containing the/ C:CH(OH) group, i.e., one in which a hydroxyl group is attached to the carbon of a double bond.

Enthalpy: Symbol H. The same of the internal energy (U) and the product of pressure (P) and volume (Nt) of a system

$$H = U + P^{V}$$

in a chemical reaction carried out at constant pressure, the change in enthalpy measured is the internal energy change plus the work done by the volume change.

$$*H = *U + P*V$$

Entropy: (Symbol S) A measure of randomness or disorder of a system. It can be defined numerically by the equation

$$ds = dq/T$$

where q is heat transferred reversible and *ds* is the entropy change for an infinitesimal reversible process.

Enzymes: The proteins which act as catalyst in biochemical reactions.

Ephedrine [$C_6HsCHOH\ CH(CH_3)NHCH_3$]: A white crystalline optically active alkaloid, m.p. 40°C, used in medicines to treat asthma, colds etc.

Epimerism: A kind of isomerism in which the isomer differs in the position of —OH groups.

Epinephrine: 2-(Nmethy-lamino)-I-(3, 4-dihydroxy-phenyl) ethanol; a chemical mediator for nerve stimulus in central nervous system. It is a hormone produced by adrenal glands that effects heart beat, blood pressure, blood sugar and causes dilation of pupil, contraction of sin blood vessels and increase in muscle blood flow.

Epoxidation: The addition of an oxygen beidge across an alkene.

Epoxyethane or Ethylene oxide: A colourless gas, soluble in water, ethanol and ether. It is obtained by passing a mixture of ethene or ethylene and air under pressure over silver catalyst at 500.

Equilibrium: A state in which the rate of forward reaction is equal to the rate of backward reaction.

Equivalent Weight: The number of part by weight of a substance which combines with or replaces 1.008 part by weight of hydrogen or eight part by weight of oxygen or 35.5 part by weight of chlorine or 108 part by weight of silver.

Erbiuve: A soft malleable silvery element of the lanthanoid series of metals. It occurs in association with other lanthanoids.

Erg: A former unit of energy used in the c.g.s. system. It is equal to 10^{-7} joule.

Ergo metrine (Ergonovine $C_{19}H_{23}N_3O_2$): A colourless crystalline alkaloid, obtained from ergot, and used in medicine to prevent haemorrhage.

Ergotamine ($C_{33}H_{35}N_5Os$): A crystalline insoluble polypeptide, m.p. 212.4°C, obtained from ergot, and used in the form of its *tartrate* in medicine as a uterine stimulant.

Ergotoxine ($C_{35}H_{41}N_5O_6$): A white crystalline insoluble alkaloid, obtained from ergot, and used in medicine as a uerine stimulant.

Ester: A type of organic compound formed by reaction between an acid and an alcohol. If the acid is a carboxylic acid, the formula is RCOOR', where rand R' are organic groups, Ester of weight acids and alcohols are volatile fragrant compounds, Esters of some bond-chain carboxylic acids occur naturally in fats and oils.

Essential Oil: A volatile oil obtained from plants and occurring especially in the flowers of a plant.

Esterification: The fatty acids react with alcohols in the presénce of like anhydrous zinc chloride or concentrated sulphuric dehydrating agent acid to form esters.

Ethanal (Acetaldehyde CH_3 CHO): A colourless, pungent smell liquid, b.p. 21 OC, miscible with water, ethanol and ether in all proportions.

Ethanamide or Acetamide (CH_3CONH_2): A colourless crystalline solid (m.p. 335 K), soluble in water, alcohol and ether.

Ethane (C_2H_6): A gaseous alkane obtained by the cracking of heavier fraction of crude oil.

1,2-Ethanediol: A colourless syrupy liquid, (b.p. 400 K) sweet in taste. It is miscible with water and ethanol. It is used as an antifreeze in automobile, preservative, as an explosive with trinitroglycerine and is a cooling agent in aeroplanes.

Ethanethiol (Ethyl Mercaptan C_2H_5SH): A colourless inflammable liquid, b.p. 37°C, used in the manufacture of rubber accelerators.

Ethanoic Acid (Acetic Acid, CH_3COOH): A pungent, corrosive m.p. 16.6°C, b.p. 118°C. It is miscible with water, ethanol and ether in all proportions. Vinegar which is 9.610 percent aqueous solution of acetic acid is obtained by the oxidation of ethyl alcohol by means of bacteria *Mycoderma aceti*. It is used as solvent and in the preparation of acetatws, acetone, acetic anhydride.

Ethanol (Ethyl Alcohol, C_2HSOH): A colourless liquid, b.p. 78.10°C. It is miscible with water in all proportions. It is obtained industrially by the fermentation of molasses or starch. It is used in the preparation of ether, ester, chloral, chloroform acid and as solvent for gum, resins, paint and varnishes and also used as a fuel.

Ethanoyl Chloride (Acetyl Chloride, CH_3COCl): A colourless fuming liquid, b.p. 52°C. It is soluble in ether and chloroform. It is mainly used as acetylating agent.

Ethene (Ethylene, C_2H_4): A colourless gas, b.p. 105°C, sparingly soluble in water. It burns with smoky luminous flame. It is prepared in laboratory by heating ethanol with excess of conc, sulphuric acid. It is used for the preparation of plastics, mustard gas, glycol, dioxan, and cellosolves.

Ethoxy Ethane (Diethyl Ether, $C_2H_5O.C_2H_5$): A colourless liquid, b.p. 34.5°C fairly soluble in water and miscible with ethanol in all proportions. It is highly inflammable and forms explosive mixture with air. It is used as solvent for oils, fats, gum, resins and used as an aneasthetic.

Ethylamine (Aminoethane, $C_2H_5NH_2$): A colourless liquid amine. It can be prepared from chloroethane heated with concentrated aqueous ammonia —$C_2H_5Cl + NH_3\ C_2H_5NH_2 + HCl$. It is used in manufacturing certain dyes.

Ethyl Carbamate [Urethan(e), $NH_2COOC_2H_5$]: A white crystalline solid, m.p. 48°C. Used in the molten state as a solvent; it is also used as an intermediate in the manufacture of resins and in medicine.

Ethyl Ethanoate (Ethyl Acetate, $CH_3COOC_2H_5$): A colourless liquid b.p. 350.5 K with a characteristic fruity odour. It is used as a solvent.

Ethyl Hydrogen Sulphate or Ethyl Sulphuric Acid ($C_2H_5HSO_4$): An oily liquid, sour in taste and having acid reaction.

Ethyl Nitrate ($C_2H_5ONO_2$): A colourless, pleasantsmelling liquid (b.p. 633.5 K). It is immiscible with water.

Ethyl Nitrite (Nitrous Ether, $C_2H_5NO_2$): A volatile liquid with a sweet smell~,b.p 17°C used in medicine.

Ethyl Orthuformate [$HC(OC_2H_5)_3$]: Obtained by running ethanol and chloroform into sodium covered with enther,

$$2CHCl_3 + 6C_2h_5OH + 6Na \rightarrow 2HC(OC_2H_5)_3 + 6NaCI + 3H_2.$$

It is used for preparing aldehydes by means of Grignard's reagent.

Ethyl Nitrate (Nitrous Ether, $C_2H_5NO_2$): A volatile liquid with a sweet smell, b.p. 17°C, used in medicine.

EthylenePropyleue Rubber (EPR): A fully saturated, stereoregular, synthetic rubber prepared by the solution polymerization of approximately equal pioportions of ethylene and propylene.

Ethyne (Acetylene, C_2H_2): A colourless gas, b.p. 840C, obtained by the action of water on calcium carbide,

$$CaC_2 + 2H_2O \rightarrow C_2H_2 + Ca(OH)_2$$

It is used for artificial ripening of fruits as an illuminant and in a oxyacetylene blow pipe for cutting and welding purposes. It is also used in manufactured acetaldehyde acetic acid, Westrol, Westron, synthetic rubber and synthetic fibre (orlon).

Euchlorine: Gaseous mixture of chlorine, Cl_2 and explosive chlorine dioxide, ClO_2.

Evaporation: A change of state from liquid to gas (vapour). It takes place at all temperature but the rate increases with temperature.

Eudiometer: A glass tube with a scale used to find out the volume of gases.

Europium: Chemical element belonging to group 3. The soft silvery white metal is used as neutron absorber in nuclear reactors. Symbol Eu;b.p. 1529°C; m.p. 822°C.

Evaporation: A change of state from liquid to gas (vapour). It takes place at all temperatures but the rate increases with temperature.

Excitation: The process of producing an excited state of an atom, molecule etc.

Excitation Energy: The energy needed to change an atom, molecule etc., from one quantum state to a state with a higher energy.

Excited State: A state of an atom, molecule or other system with an energy greater than that of the ground state.

Exothermic Reaction: A reaction that proceeds with the evolution of heat, e.g., carbon, sulphur and magnesium combined with oxygen.

F

Fajan's Method: The titration of Cl^- with Ag^+ using fluorescein as an adsorption indicator. At the end point the colour of the precipitate is red.

Fajan's Rules: The polarization of anion takes place when two oppositely charged ions approach each other. The polarisation of anion is favoured by large size of the ions, small cation and large anion, nature of the solvent and nature of the cation.

Faraday: (Symbol F) A unit of electric charge equal to the charge required to discharge one mole of a singly charged ion. One Faraday is 9.648670 x 104 coulombs.

Faraday's Laws (of electrolysis): *First Law : The* amount of substances decomposed as a result of electrolysis is directly proportional to the quantity of electricity passed

$$w \propto c \times t$$

where w is the weight of the substance decomposed, c is the current strength in amperes in time *t* second.

Second Law : The amount of different substances liberated by same quantity of electricity proportional to their equivalent weights.

Faraday Unit: The electrical charge of one mole of electrons, I Faraday = 96487 coulombs.

Fehling's Solution: A solution used to detect the presence of aldehydic group (—CHO). It is a mixture of freshly prepared copper (II) sulphate solution with alkaline potassium sodium 2, 3dihydroxy butanedioate (tartrate). The aldchydic group gives red precipitate on heating with Fehling's solution due to the formation of cuprous oxide.

Feldspars: A group of aluminosilicates with the framework structures with Al or Si in tetrahedral coordination. It is used in the ceramic and enamel industries.

Femto: Symbol + prefix denoting 10^{15}. For example, femto *(fin)* = 10^{15} metre (m).

Femto: *(f)* Prefix denoting 10^{-15} ; i.e. one femtometre *(fm)* is equal to 10^{-15} metre.

Fermentation: A chemical change brought about in organic substances by the enzymatic action of living organisms such as yeast and bacteria, *e.g.,* fermentation of sugars

$$C_6H_{12}O_6 \rightarrow 2C_2H_5OH + 2CO_2$$

Fermi: A unit of length equal to 10^{-15} metre. It was formerly used in atomic and nuclear to physics.

Fermiuve: A radioactive transuranic element of the actinoid series, not found naturally on Earth. Several short lived isotopes have been synthesized.

Ferrocene [Fe(C$_5$H$_5$)$_2$]: An orange crystalline solid. It is an example of a bondwich compound in which an iron (11) ion is coordinate to two cyclopentadienyl ions. The boding involves overlap of orbitals on the iron with the electrons in the cyclopentadienyl ring. The compound melts at 173°C, is soluble in organic solvents, and can undergo substitution reactions on the rings. It can be oxidised to the blue cation $(C_5H_5)2F_e^{+}$.

Ferromagnetism A type of magnetism in which magnetization persists even after the applied field is removed. Iron nickel and cobalt are ferromagnetic.

Fibre Reactive Dyestuffs: Reactive dyestuffs that form a direct chemical bond to the OH groups of cellulose fibres and thus form very resistant colours.

Fick's Law of Diffusion: A law linked to the rate of diffusion of a substance in a given direction to the gradient of its concentration.

$$\text{Thus, } NA = -D_{AB} \frac{dC}{dz}$$

where NA is the molar rate of diffusion/unit area. DAB' is the diffusivity of substance, A in substance B; C_A is the molar concentration of A; z is the distance in the direction of diffusion.

Fieser's Solution: An aqueous alkaline solution of sodium anthroquinone b-sulphonate (silver salt) reduced with sodium dithionite, Na_2S2O_4. It is used as a scrubbing solution for partially removing O_2 from say N_2.

Fertilizers, Multinutrient: The chemical fertilizers which can supply more than one nutrient to plants, e.g., ammonium phosphate (N and P); potassium nitrate (K and N), ammonium superphosphate (N, Ca and P) etc.

Fertilizers, Nitrogenous: Those fertilizers which give nitrogen to p ants in readily available form, *e.g.*, ammonia (NH_3, 82% N), ammonium nitrate (NH_4NO_3, 33% N), Ammophos or ammonium phosphate ($NH_4H_2PO_4$; $(NH_4)2HPO_4$,I6% N + 20% P_2O_5), ammonium sulphate ($(NH_4)2SO_4$,20.6% N), calcium ammonium nitrate ($CaCO_3$ + NH_4NO_3, 20.5% N), calcium cyanamide (Ca NCN, 15.5% N), potassium nitrate (KNO_3,15% N, 44% K20) and Urea (H_2N $CONH_2$,44-46% N). Among these, urea is the largest used nitrogenous fertilizer.

Fertilizers, Phosphatic: The compounds which can supply phosphorus to plants in a readily available form, e.g., calcium superphosphate [Ca $(H_2PO_4)2.CaSO_4$, single super phosphate, 16% P_2O_5; triple superphosphate, 48% P_2O_5), ammonium super phosphate ($NH_4H_2PO_4$ + $CaBPO_4$ + $Ca_3(PO_4)_2$ + $(NH_4)_2SO_4$,16-18% P_2O_5 + 34% N), calcium metaphosphate $Ca_3(PO_3)2$, 62-63% P_2O_5), raw bone meal (20% P_2O_5) and basic slag (23-30% P_2O_5 + 3% N).

Fertilizers, Potassic: The compounds which can supply readily available potassium to soil and plants, *e.g.*, potassium chloride (KC1, 48-60% K_2O), potassium nitrate (KNO_3, 44% K_2O, 10% N), potassium sulphate (K_2SO_4, 48-50% K_2)) and wood ash (mostly K_2CO_3 + SiO_2, 3-7% + 1-2% P_2O_5).

Filler: A solid material used to modify the physical properties or reduce the cost of synthetic compounds. Such as rubbers, plastics, paints and resins. State powder, glass fibre, mica, and cotton wool are all used as fillers.

Film Badge: A badge containing marked photographic film worn by workers likely to be exposed to ionizing radiation to indicate the extent of their exposure to their harmful radiations.

Filtration: A process of separating insoluble solid particles from a liquid by passing the fluid through porous material (filter paper),

Filter Pump: A type of vacuum pump in which a jet of' water forced through a nozzle carries air molecules out of the system. Filter pumps cannot produce pressures below die vapour pressure of water. They are used in laboratory for vacuum filtration, distillation and techniques requiring a low grade vacuum.

Fine Organic Chemicals: Carbon compounds, such as pesticides, dyes, and drys, produced only in small quantities. Their main requirement is that they must have a high degree of purity, often higher than 95%. They are manufactured for special purposes, e.g., for use in spectroscopy, pharmacology and electronics.

Fire Extinguisher: An apparatus which supplies a fluid which stops combustion, e.g., an extinguisher using carbon dioxide foam.

Fire Retardant: An additive that causes resistance to burning when in contact with highenergy ignition source

First Order Reaction: A reaction in which the rate of reaction is proportional to the concentration of one of the reacting substances, for example, hydrolysis of ester.

Fischer-Tropsch Process: A process in which water gas mixed with half its volume of hydrogen is then passed over the catalyst at 470-870 K under 110 atom pressure to get a mixture of hydrocarbons.

Fitting Reaction: The production of a benzene homologue from an aromatic halide by the action of metallic sodium in ether.

Five Structure: Closely places lines seen at high resolution in a spectral line or bond. Five structure may be caused by vibration of the molecules or by electron spin. Hyperfine structure, seen at very high resolution, is caused by the atomic.

Fixed Oil: A nonvolatile oil occurring in plants. It is used in cooking, e.g., peanut oil and coconut.

Flame Photometry: A development of the flame test used in qualitative analysis; photometric measurement of flame emission is used to find out the concentration of substances introduced into the flame,

Flash Photolysis: A method used to study very fast reaction that, usually, involves atoms or radicals in the gas phase.

Flash Point: The lowest temperature at which sufficient vapour is given off by a flammable liquid to ignite in the presence of spark.

Flame Test: Some basic radicals in the form if their chlorides impart characteristic colour to the nonluminous flame. The underlying principle involved is that the chlcrides are thermally ionized. The ion so obtained absorbs heal and gets excited. The extra energy is given out in the form of light of a characteristic colour.

Flare Stack: A chemistry at the top of which unwanted gases are bound.

Flint: Natural variety of impure silica, SiO_2. 'Flints' of automatic lighters are composed of pyrophoric alloys of metal such as cerium and iron.

Flint Glass: A type of glass having lead silicate. It is used for optical purposes.

Flocculating Agent: A substance which forms a gelatinous precipitate on addition to water. The precipitate settles in the water and carries with it finely divided particulate matter which does not settle out of its own accord.

Fluoridation: The introduction of small quantifies of flouride compounds into the water supply as a publichealth measure to reduce the incidence of booth decay in children.

Fluxional Molecule: A molecule in which the constituent atoms change their relative positions as quickly at room temperature that the normal concept of structure is inadequate, i.e., no specific structure exists for longer that about 10^{-2} second and the relative positions become indistirequistable. For example, CIF_3 at –60°C has a distinct 'T' shape but at room temperature the fluorine atoms are visualized as moving rapidly over the surface of the chlorine atom in a state of exchange and are effectively identical.

Fuel Gas: The gaseous products of combustion from a boiler furnace consisting predominantly of carbon dioxide, carbon monoxide, oxygen, nitrogen and steam. Analysis of the flue gases is used to check the efficiency of the furnace.

Fluidization: A technique used in industrial chemistry, in which a mass of solid particles is brought into a state of suspension by an upward stream of gas blown through it in a reactor. The material in the resultant 'fluidized bed', which resembles a boiling liquid, is more accessible to, chemical reaction, than the same solid material in the static state.

Fluorescein ($C_{20}H_{12}O_5$): A dark red crystalline organic compound, m.p, 314°C. It dissolves in alkaline solutions to give a liquid of intense green fluorescence. It is used in dyes as an indicator.

Fluorescence: A property of a substance to emit electromagnetic radiations by absorbing energy.

Fluorimetry: An analytical method using the fluorescent properties of the substance to be estimated.

Fluorinating Agent: A substance used to introduce fluorine or replace it.

Fluorine: A pale greenish yellow gas belonging to group 17. It is prepared by the electrolysis of a solution potassium hydrogen fluoride in anhydrous hydrogen fluoride. It is the most reactive of all the elements. It combines directly with almost all other elements with extreme vigour forming fluorides.

It is used to prepare freon which is used in refrigeration and as solvent for D.D.T. Its compounds, such as, sodium fluoride and cryolite are used as insecticides. A new plastic teflon which is resistant to the action of acids, alkalies and oxidizing agents, is a polymerised tetrafluorethylene ($F_2C = CF_2$). Uranium hexafluoride (UF_6) is used in separation of isotopes of uranium. It can prevent tooth decay if added to drinking water.

Fluorocarbon: A compound derived from a hydrocarbon by replacing hydrogen atom with fluorine, e.g., from ($CC1_2H_2$).

Fluorosilicic Acid: (Hydrofluosilicic Acid, H_2SiF_6) An acid that is only stable in the form of its fuming aqueous solution. Used as a disinfectant and wood preservative.

Fluorspar: A naturally occurring form of calcium fluoride, CaF_2.

Flux: A material which combines with the gunge and forms an easily fusible material called slag which is not soluble in the molten metal. e.g.,

$$CaO + SiO_2 \rightarrow CaSiO_3$$

Fluxinal Molecule: A molecule in which the constituent atoms change their relative positions so quickly at room temperature that the normal concept of structure is inadequate, i.e., no specific structure exists for longer than about 10^{-2} second and the relative positions become indistinguishable. For example in chlorine trifluoride, the molecule CIF_3 has a distinct T-shape at 60°C but at room temperature the fluorine atoms are visualized as moving rapidly around chlorine atoms in a state of exchange and are effectively identical.

Fly Ash: Very fine powdery noncombustible residue left after turning of coal in a thermal power station. It is used as an essential ingredient of Pozzolona cement.

Foam: A dispersion of air or gas bubbles in a liquid, as soap suds. Used as cushioning and insulating materials, liquid foam is used to extinguish fires.

Food Chain: Transfer of food energy from plant source back into plants through a series of organisms with repeated eating, being eaten and final mineralization. It is a way of describing how all animals depend on each other for food.

Food Value: The energy released when a fuel or food is combusted.

Force Constant: A measure of the restoring force to vibration in a bond.

Formality (Symbol F): The number of gram—formula weights in a litre of solution. It is similar to morality and its use avoids the difficulty of assigning a molecular weight to something (such as NaCI) that contains no discernible molecules. I-F solution of NaCI is prepared by dissolving 58.5 gm of NaCI in one litre of solution.

Formula: A representation of a chemical compound in symbolic form of its constituent elements together with their number placed below and to the right of each symbol.

Formulation: An economic, efficient and safe combination of toxicant(s), adjuvant(s) and diluent(s) (base material or fillers) used for ultimate control of insect pests or diseases, It may be any treatment of the active ingredient that improves its application, effectiveness, handling, storage and interaction with the ecosystem, i.e., the formulation yields the form in which pesticide/medicine is prepared for practical use, e.g., wettable powder, tablet, dust etc.

Fraction: A mixture of liquids with similar boiling points collected by fractional distillation.

Fractional Crystallisation: A process of separating a less soluble substance from a more soluble substance from a solution. If solution contains two substances in equal amounts, the one with lower solubility crystallizes out first. In case a solution contains two substances whose solubilities are nearly equal at some temperature, then the one with greater amount crystallizes out first.

Fractional Distillation: A process of separating two or more immiscible liquids having different boiling points. It is achieved by distillation with a fractionating column.

Fractional Surface Coverage: (Symbol q) The fraction of all sites on a solid surface which are occupied by adsorbed molecules.

Frasch Process: A process to obtain sulphur. The sulphur is obtained in large scale from underground beds by melting it with water heated at 170'(" A froth of sulphur, air and water is forced to the surface by hot, compressed air. The sulphur, thus, formed is about 99.5% pure.

Free Electron: An electron that is not attached to an atom, molecule, or ion, but is free to move under the influence of an electric field.

Free Energy: Gibbs Function, G. A thermodynamic quantity representing the energy that would be liberated or absorbed during a reversible process. Defined, under conditions of constant temperature and pressure, by $G = H - TS$, where H is the heat content (enthalpy), T the thermodynamic temperature, and S the entropy.

Free Radical: An atom or group of atoms with a single impaired electron. It is produced by breaking a covalent bond in symmetrical fashion.

Freezing: The process by which a liquid is converted into a solid by cooling the reverse of melting.

Freezing Mixtures Two or more substances mixed together to produce at low temperature. A mixture of sodium chloride and ion in water (–20°C) is a common example.

Freezing Point The temperature at which a liquid is in equilibrium with a solid phase at standard pressure and below which the liquid freezes or solidifies.

Freons: A mixture of difluoro-dichloro-methane (b.p. 28°), 1, 1-dichloro-1, 2, 2, 2-tetrafluoroethane (b.p. 2°C) and 1, 2-dichloro-1, 1, 2, 2-tetrafluoroethane (b.p. 4°C) commonly used propellent for aerosols and refrigeration liquid.

Friedl's Craft Reaction: A method for the substitution of an alkyl or acyl group on to a benzene in the presence of anhydrous aluminium chloride (catalyst).

Fries Rule: The most stable arrangement of double bonds in a poynuclear compound in the one in which the maximum number of rings possess the benzenoid structure (having three double bond).

Froth Flotation: A method to separate ore mineral from gunge by mixing with water and little of vegetable oil. Compressed air is circulated and froth formed, taking away the ore particles. This method is used to concentrated sulphide ore.

Fructose ($C_6H_{12}O_6$): A white crystalline solid (m.p. 375 K with decomposition). It is the sweetest of all sugars. It is found in fruits and honey. It is used as sweetening agent.

Fuel Cell: The electric cell designed to convert the energy from the combustion of fuels (such as, hydrogen or methane) directly into energy. A fuel cell, such as that used on space craft and designed to use the reaction of II_2 and O_2 to form water. Hydrogen under positive pressure is diffused through a porous graphite anode impregnated with a finely powdered platinum or palladium catalyst. Oxygen is similarly diffused through a porous graphite cathode impregnated with finely divided cobalt (II) oxide, CoO, platinum, or silver as a catalyst. Both the electrodes are immersed in the same concentrated hydroxide ion solution. The efficiency of a fuel cell is about 70%. Fuel

Element: An element of nuclear fuel for use in a nuclear reactor, usually uranium encased in a can.

Fullerenes: A newly discovered family of the third allotropic form of carbon. The first member of the family was a molecule made of 60 carbon atoms shaped like a geodericdome. It was called buckinster fullerene. Other similar molecules have now been discovered like C_{28}, C_{32}, C_{70}. There are called fullerenes.

Fuller's Earth: A variety of claylike materials that absorb oil and grease. They consist of hydrated silicates of magnesium, calcium, aluminium, and sometimes other metals. Used in scouring textiles and in refining fats and oils.

Fumaric Acid ($C_4H_4O_4$): A colourless crystalline solid (m.p. 560 K) soluble in water.

Fuming Nitric Acid: A brown fuming highly corrosive liquid consisting of nitric acid containing an excess of nitrogen peroxide, Used as an oxidant in rockets and in organic synthesis.

Functional Group A group of atoms in an organic compound that is responsible for their chemical properties, e.g., aldehyde —CHO, alcohol —OH, amine NH2, ketone Carboxylic acid —COOH, etc.

Functional Isomerism: Structural isomerism due to difference in functional groups, e.g., ethanol (CH_3 CH_2OH) and dimethyl ether (CH_3OCH_3) are isomeric. Same is the case with acetone (CH_3COCH_3) and propanal (CH_3CH_2CHO).

Fundamental Units: The units of length, mass and time that form the basis of most systems of units. In SI, the fundamental units are the metre, the biloogram, and the second.

Fungicide: A substance capable of destroying harmful fungi, such as, moulds and mildews.

Fungicides, Azole: Fungicidal derivatives of 1, 3-diazole; benzimidazole; 1,2, 4-triazole; isooxazole and thiazole. These compounds are wide spectrum fungicides with systemic and therapeutic action against a number of fungal species, They inhibit DNA/RNA synthesis or ergosterol synthesis in fungal spores.

Fungicides, Carboximide: (Oxathin Fungicides) A group of fungicides that is derived from 1, 4 oxathin, e.g., 5, 6-dihydro-2-methyl-1, 4-oxathiin-3-carboxanilide (carboxin) and 5, 6-dihydro-2-methyl-I, 4 oxathiin-3-carboxanilide-4, 4-dioxide (oxycarboxin). These systemics have selective chemotherapeutic action against *Basidomycetes* fungi, e.g., smuts and rusts, as well as against *Rhizoctonia* solant. Active carboximides inhibit acetate, pyruvate and succinate synthesis in fungi and inactive succinate dehydrogenase.

Fungicides, Dithiocarbamate: Esters and salts of N-alkyl dithiocarbamic acid, e.g., $RNHC(S)SCH_3$ or RN HC(S)SM; and N,N-dialky ldithiocarbamic acid R1R2 $NC(S)SCH_3$ or $R_1R_2NC(S)$ SM where R, R_1, R_2 are alkyl groups and M is a metal ion. Some popular compounds of this group are: sodium methyl dithiocarbamate, vapam; zinc N, N-diisopropy ldithiocarbamate, neviram; zinc N, N ethylenebis dithiocarbarnate, zineb; manganese N, No-ethylene bisdithiocarbwnate, maneb; and mixtures of polyethy lenebisthiuram sulphides and its zinc, complex, metiram. These compounds are widely used for control of fungi on fruits in orchards. The dithiocarbamates react with SH group containing enzymes and coenzymes of fungal cells.

Fungicides, Elementorganic (Organometallic Fungicides): Alkyl or aryl metal salts of mercury (R-Hg-V,, and tin (R_3SnX) used as fungicides, *e.g.* Methoxymethyl murcury acetate and triphenyl tin acetate.

Fungicides, Haloaromatic: A group of fungicides that is derived by halogenation of aromatic compounds, inclu de chloronitrophenylethers, chlorophenols, nitrochloro berizenes and esters of nitro phenols and the chloroqui nones.

Fungicides, N-Mercaptoamide: N-sulfenyl alkyl derivatives of amides of carboxylic acids or sulphonic acids, These are protective foliage fungicides, with a broad spectrum of activity.

Fungicides, Organophosphorus: A group of organophos thates, thiophosphates, di-thiophosphates and their amides, known to possess fungicidal properties.

Furan or Oxacyclopenta-2, 4 diene: A colourless liquid (b.p. 305 K) obtained by distillation of wood especially pinewood.

Furanose: A sugar that has a five-membered ring (four carbon atoms and one hydrogen atom).

Furfural or Furfuraidehyde or 2-Furaidehyde ($C_5H_4O_2$): A colourless liquid with a peculiar odour (b.p. 162°C).

Furfuryl Alcohol ($C_5H_6O_2$): A colourless liquid (b.p. 170-171°C) obtained by the reduction of furfural or with furoic acid by the I action of 30% NaOH solution on furfural in cold. It is poisonous.

Fused: Describing a solid that has been melted and fossolidifies into a single mass. Fused silica, for example, is produced by melting bond.

G

Galactose ($C_6H_{12}O_6$): A sugar found in lactose and many polysaccharides. It may be prepared by hydrolysing lactose, It behaves like glucose.

Gallic Acid or 3, 4, 5-Trihy-droxy Benzoic Acid ($C_7H_6O_5$): A colourless crystal with the one molecule of water (m.p. 253°C), sparingly soluble in water and alcohol. It is found free in woody tissue, in gallnuts and in tea, and is a constituent of the tannins, from which it can be obtained by fermentation or by acid hydrolysis. It is used in the manufacture of inks.

Gallium: A silver white metal belonging to group 13. Germanite is the only mineral which contains gallium. It is used in lamps for spectroscopic analysis.

Symbol : Ga; am. 3 1; m.p. 29.78°C; b.p. 2403**Galactose (C_6H12O6):** A sugar found in lactose and many polysaccharides. It may be prepared by hydrolysing lactose, It behaves like glucose.

Galvanizing: A process of covering steel with zinc by dipping in a vessel of molten zinc or by electro deposition. This is done to protect steel from corrosion.

Gamma Rays (Gamma Radiation, y-Rays): Electromagnetic radiation of the same nature, but shorter wavelength than X-rays (10^{-10} -10^{-13} metre). They are emitted by the nuclei of radioactive atoms during decay. Gamma rays are emitted in quantized units called photons.

Gangue: A worthless matter, like sand and rock which forms part of a mineral.

Garnet: A group of minerals of varying composition, mainly double silicates of calcium or aluminium with other metals. Several varieties are red in colour, and are used as gems.

Gas: The state of matter in which forces of attraction between the particles of a substance are less. The gas particles can expand indefinitely, be compressed readily, freely mix with other gases, and show low density. They have no fixed shape or volume.

Gas Chromatography: A technique widely used for the separation and analysis of mixtures.

Gas Constant (Symbol R): The universal constant 8.31434 J $mol^{-1}k^{-1}$ representing the proportionally between pressure, volume, and temperature at which a gas exists.

Gaseous Diffusion Separations: A technique for the separation of gases that relies on their slightly of different atomic masses.

Gas Laws: Laws relating the temperature, pressure and volume of a fixed mass of gas, The main gas laws are Boyl's law and Charle's law. The laws are riot obeyed exactly by any real gas, but many common gases obey them under certain conditions, particularly at high temperatures and law pressures. A gas that obeys the laws over all pressures and temperatures in a perfect or ideal gas. Boyl's and Charle's laws can be combined into an equation of state for ideal gases

$$PV_m = RT$$

where V_m is the molar volume and R the molar gas constant. For in moles of gas $PV = nRT$.

Gas Liquid Chromatography: This is gas chromatography where a column packed with a solid coated with a nonvolatile liquid is employed.

Gas Mantle: A structure composed of the oxides of thoriurn (99%) and cerium (1%) made by impregnating a combustible fabric with a solution of the nitrates of the metals, and decomposing the nitrates by beat.

Gas Mask (Respirator): A device for protecting the face and breathing organs against poisonous 'gases'. (These include poisonous smokes, etc., used in chemical warfare.) The air is drawn through a layer of activated carbon, which adsorbs vapours, and also through a filterpad, which retains solid particles of smokes. Such an arrangement is effective against war 'gases' and smokes, but not against gases of low molecular wet. ght, such as carbon monoxide or coalgas.

Gas Solid Chromatography: A column packed with a solid stationary phase is employed for this type of gas chromatography.

Gattermann Reaction: The combination of phenol with hydrogen chloride or hydrogen cyanide to give a product which is hydrolysed by water to an aldehyde.

Gay-Lussac's Law: In any reaction which involves gaseous reactants or products, there always exists a simple whole number ratio between the volume of these gases if all measurements of volumes are being made under similar conditions of temperature and pressure.

Geiger Counter: An instrument which tests for radioactivity and measures, the strength of alpha, beta or gamma rays,

Gel: A lyophilic colloid that is normally stable but may be induced to coagulate partially under certain conditions. This produces a pseudosolid or easily deformable jelly like mass, called a gel, in which intertwining particles enclose the whole dispersing medium. Gels may be further subdivided into elastic gels and rigial gels.

Gel Filtration: This is modified form of molecular sieve. The sieve is a continuous gel mode form a polysaccharide. In this case, molecules larger than the largest pore size are totally excluded from the column.

Gem Positions: Positions in a molecule on the same atom. For example, 1, 1-dichloroethane (CH_3CHCl_2) is a gem dihalide.

Germane (Germanium Tetrahydride GeH_4): A colourless gas. The germanium analogue of methane; the first member of a series of organogermanium compounds of the general formula Gen H2n+2, corresponding to the alkanes.

Germanium: A silver white element belonging to group 14. It is hard and brittle. It is more, reactive than silicon. It is used as semiconductor especially in transistors.

Symbol : Ga; a.n.32; m/p. 937.4°; b.p. 2830°C;r.d. 5.32

Germanium (IV) Oxide (GeO_2): A compound made by strongly heating germanium in air or by hydrolysing germanium (IV) chloride. It occurs in two forms. One is slightly soluble in water forming an acidic solution and the other is insoluble in water. Germanium oxide dissolves in 2 alkalis to form the anion $GeO3^{-2}$.

g-Factor: The proportionally factor in the relation between the magnetic moment m and the number of unpaired electrons

$$\mu = g \sqrt{s(s+1)}$$

For a free electron $g = 2.003$.

Gibberellins: Gibberellic acid (GA_3) and its derivatives, having plant growth regulatory action; generally used for reduction of russetting in certain apple cultivars, to check crown rot in banana and bunch rot of grapes: increasing berry size and elongation of clusters. Physiological effects of gibberellins include production of internode and hypocotyl growth seed germination; production of flowering, bolting and their development, cell division in the cambial cells and increased enzyme, activity.

Gibbs Function (G): A thermodynamic function defined by G = H–TS, where H is enthalpy, T the thermodynamic temperature and S the entropy. G is minimum for conditions of chemical equilibrium for reactions at constant temperature and pressure.

Giga: A prefix denoting 10^9, i.e. one gigametre (Gm) is equal to 10^9 metre; one gigahertz (GHz) is equal to 10^9 Hertz.

Glass: A hard, brittle amorphous substance, usually transparent or translucent comprising of a mixture of silicates of Na, Ca or other metals. It is made by melting together sand (silica), sodium carbonate and lime. Glass used for optical instruments contains lead, potassium or barium. It softens when heated at a certain temperature. It is unaffected by water, acid etc., except hydrofluoric acid which dissolve it with the liberation of silicon tetrafluoride gas.

Glucose (Dextrose, Grape Sugar, $C_6H_{12}O_6$): A monosaccharide occurring widely in nature as D-glucose. It occurs as glucose units is sucrose, starch and cellulose. It is important to metabolism as it participates energy storage and energy released systems.

Glutamic Acid: A colourless crystalline amino acid, m.p. 206⁰C, used in the form of its sodium salt as a flavouring agent.

Glutamine Conjugation: Conjugate formed with amino group of glutamic acid (in Proteins) with a herbicide of phenyl acetic acid type resulting in formation of an amide.

Glyceride: An ester formed between glycerol (propane-1, 2, 3, -trid) and one or more carboxylic acids, Glycerol has three alcohol groups and if all three form ester die compound is a triglycerides, Naturally occurring fats and oils are B-triglycerides, of longchain carboxylic acids (hence the name fatty acid).

Glycogen (Animal Starch): A polysaccharide that the main carbohydrate store of animals. It is composed of many glucose units linked in a similar way to starch. Glycogen is readily hydrogen a stepwise manner to glucose its of. It is stored largely in the liver and in muscle but is from widely districted.

Glycine (Amino Acetic Acid, Glycoll: A white solid found in many proteins and occurs in certain animal excretions. It is prepared by the action of concentrated ammonium hydroxide solution on chloroacetic acid.

Glyoxal [Diformyl $(CHO)_2$]: A yellow crystalline substance, m.p.150C, b.p. 51°C. Used in the manufacture of plastics, and in textile finishing.

Glyptals or Alkydes: Manufactured by the action of pthalic acid or its anhydride or other dicarboxylic acids on glycerol. It is mainly used as coating in lacquer and paint industry.

Gold: A yellow coloured transition metal belonging to group II. It is sufficiently hard but softer than copper and silver. It is highly malleable and ductile. It is a noble metal having no action of air, moisture, caustic alkalies and acids. It dissolved in aqua regia. It is used in making ornament, coins and employed in photography for toning. It is used for goldplating of articles of baser metals.

Symbol : Au; a.n. 79; m.p. 1063°C; b.p. 2710°C; r.d. 19.3.

Gold (III) Chloride (Auric Cutoride, $AuC1_3$): A brown mass, deliquescent and readily dissolves in water. It is used in electroplating and in photography as a toning agent.

Goldschmidt Process: A process for extracting certain metals from their oxides by reduction with aluminium.

Grain: A crystal in a metal that has been prevented from attaining its regular geometrical form,

Granite: A very hard igneous rock containing at least three minerals mica, felspar and quartz.

Granulation: A process of enlarging particles to improve the flow properties of solid reactions and products in chemical processes. The larger a particle, and the freer from time materials in a solid, the more easily will it flow. Dry granulation pollets from dry materials, which are crushed into the desired size. Wet granulation involves the addition of a liquid to the material, and the resulting paste is extruded and dried before cutting into the required size.

Graphite: A crystalline solid in the form of hexagonal plates having a greasy feel. It is an allotropic form of carbon. It is a good conductor of heat and electricity due feel to electrons. It is used as lubricant and in the preparation of electrodes

Gravimetric Analysis (Quantitative Analysis): A process of isolating and weighing an element or a definite compound on the element in as pure a form as possible. The steps which determine a successful analysis by precipitation are : (1) The substance to be analysed is converted into an insoluble, Substance; (2) The physical separation of the precipitate by filteration, (3) The precipitate is converted into a pure substance by ignition or by a simple, operation,

Gray: (Symbol Gy) The SI unit of absorbed energy dose per unit mass resulting from die passage of ionizing radiation through living tissue.

Grignard's Reagent: A colourless nonvolatile solid obtained by the interaction of dry magnesium turning with alkyl halide in the presence of dry alcohol of free ehter

$$RX + Mg \rightarrow RMg\,X$$

it is used in the preparation of alcohol, thioalcohol, carboxylic acids, ether, aldehydes, ketones, esters etc.

Ground State: The lowest energy state of an atom, molecule or ether system; it is the most stable state.

Group: The elements in a column of the periodic table having similar electronic configurations, *e.g.,* alkali metals having electronic configuration nS1.

Gun Cotton: Cellulose nitrate, nitrocellulose. A powerful explosive formed by the action of nitric acid on cellulose.

Gunmetal: A variety of bronze containing about 99% copper, 8%, 10% tin, and up to 4% zinc. Formerly widely used to make cannons because of its ability to withstand pressure and heat.

Gun Powder: A powdered mixture of sulphur, charcoal and potassium nitrate. It is used as an explosive and in foreworks.

Gypsum: Natural hydrated calcium sulphate ($CaSO_4$ $2H_2O$). It is a white to yellowish-white mineral. When heated to 120°C, it loses ¾ of its water and forms plaster of paris.

H

Haber's Process An imporatant industrial process for the preparation of ammonia, which is used for fertilizers and for making nitric acid. the reaction is the equilibrium:

$$N_2 + 3H_2 = 24H_3 + heat$$

To get good yield of NH_3 high pressure (250 atm), low temperature (400°C) and a catalyst ($Fe + Al_2O_3$) is required.

Hadron A major group of subatomic particles which interact strongly) with one another. They include protons, neutrons and quarks.

Haematite (Fe_2O_3) A mineral form of iron (iii) oxide. It is the main ore of iron.

Haemoglobin: The frigment of the red blood cells that is responsible for the transport of oxygen from lungs to the tissues. It consists of a basic protein, globin, linked with four haem groups. Haem is a complex compound containing an iron atom.

Hafnium: A white metal belonging to group 4. It bums in air to form HfO_2. The metal is slowly attacked by acids except HF and aqua regia. It is used in making wireless valves and in surgery.

Symbol Hf, am. 72; m.p. 2220^0 C; b.p. 5230°C; r.d. 13.3.

Half Cell: Half of an electrolytic cell, consisting of an electrode dipping into an electrolyte. The electrode potential of such a system is measured by comparison with a hydrogen electrode, which is assigned an electrode potential of zero.

HalfLife: The time required for half of the radioactive isotope in any given sample to disintegrate.

HalibutLiver Oil: Obtained from the liver of the halibut, It is rich vitamin A and used in the treatment of vitamin A deficiency diseases.

Haloform: The organic compounds with general formula CHX_3, where X is a halogen atom (Flourine (F), chlorine (CI), bromine (Br), iodide (I)].

Halogenation: A reaction involving addition or substitution by halogen atom in a molecule, *e.g.,* halogenation of methane,

$$CH_4 + Cl_2 \rightarrow CH_{3Cl} + HCl$$

addition HI in ethylene.

$$CH_2 = CH_2 + HI \rightarrow CH_3CH_2I$$

Halogens: The five elements flourine (F), chlorine (CI), bromine (Br), iodide (1) and astatine (At) belonging to group 17.

Halonium Ions: Disubstituted halogen cations, RXR', where RR' may be alkyl or aryl groups. They are used as alkylating agents.

Halothane or Fluothane ($CH\ BrCl.CF_3$): Colourless, mobile heavy liquid with an odour resembling chloroform. It is noninflammable, b.p. 49-51°C. It is a general anesthetic administered in oxygen often along with dinitrogen oxide.

Hamoner Mills: Devices used in the chemical industry for crushing and grinding solid materials at high speeds to a specified size. The impact between the particles, grinding plates and grinding hammers pulverises the particles.

Hantzsch Synthesis: The formation of pyridine derivative by condensation of ethyl acetoacetate with ammonia and an aldehyde.

Hardness: The resistance of a material to pressure applied to a small area, *i.e..* resistance to abrasion, crushing, indentation or streching.

Hardness of Water: The water which gives curdy precipitate with soap solution, It is due to the presence of bicarbonate, chloride and sulphate of magnesium and calcium,

Hardening of Oils: A process for converting liquid vegetable oils into solid or semisolid fats. Oils axe glycerides of mainly unsaturated carboxylic acids. Hydrogenation of the double carboncarbon bonds using a middle catalyst produces saturated glycerides which have higher melting points.

Hard Water: Water which does not form leather with soap due, to the presence of Ca, Mg and Fe compounds. Removal of these compounds renders the water soft,

Hartree (Atomic unit of energy): A unit of energy equal to e^2/ao, where e is the charge of an electron and ao is the atomic unit of length. It is equal to 4.85×10^{-18} joule or 27.2 eV.

Heat of Atomization: The heat required in dissociating one mole of a substance into atoms.

Heat Capacity: The amount of heat required to raise one mole of material, one celsius degree.

Heat Engine: A devices for converting hot energy into work. Heat engines operate by transferring energy from a high temperature source to a low temperature sick. The theoretical operation of heat engines is useful in the theory of thermodynamics.

Heat Exchangers: Devices that enable the heat form a hot fluid to be transferred to a cool fluid without allowing them to come into contact. The normal arrangements is for one of the fluids to flow in a coiled tube through a jacket containing the second fluid. Both the cooling and heating effect may be of benefit in conserving energy used in a chemical plant and in controlling the process.

Heat of Combustion: The heat liberated when one mole of a substance burps in excess oxygen.

Heat of Crystallisation: The amount of heat evolved when a unit mass of a liquid crystallizes at its freezing point.

Heat of Dissociation: The energy required to dissociate one mole of a substance into its constituent elements.

Heat of Formation: The energy change when one mole of a substance is formed from it, element.

Heat of Neutralization: The heat liberated when one mole of an acid or base is neutralized.

Heat of Reaction: The amount of heat changed when molar amount of a given substance reacts completely.

Heat of Solution: The amount of heat changed when one mole of a substance is dissolved in a given solvent to infinitedilution.

Heat Transmission Oil: Thermally stable oils usually Indium viscosity distillates employed as circulated heat transfer media for industrial purposes.

Heat of Vapourization: The quantity of heat required to vaporize a unit mass of a given liquid at constant temperature.

Heavy Water (Deuterium Oxide, D_2O,. The compound obtained by exchanging deuterium and hydrogen atom in water

$$H_2O + D_2 = D_2O + H_2$$

It is used as a tracer in metabolic studies and also used as moderator in nuclear reactors.

Symbol: b.p. 101.4°C; m.p. 3.8°C.

Hecto Prefix for denoting 102 or one hundred times, *i.e.* one hectometre is one hundred metre.

Height Equivalent to Theoretical Plate (HETP): A measure of column efficiency in chromatography calculated as equal to L/N, where L is the length of the column and N the number of theoretical plates. The number of plates, $N = 16(x/y)^2$ where, x is distance of injection peak (base line) to peak maxima and y is the base width of the peak.

Heisenberg's Uncertainty Principle: Werner Heisenberg showed that it is impossible to measure simultaneously the exact position and exact velocity of a body as small as an electron.

Heliocides: The terpenoids derived from hemigosspolone and possessing insecticidal activity. These are present in flower buds of some primitive and wild strains of cotton which show resistance to bollworm and tobacco bud worm. Pour heliocides H_1, H_2, H_3 and H_4 have been reported.

Helium: A not inflammable, mono atomic gas belonging to group 18, obtained from liquid air or natural gas. It is used for filling weather observation baloons, in 'neon signs' and cooling of electric equipment in the treatment of respiratory diseases.

Symbol : He; a.n. 2; m.p. 272.1°C; b.p. 268.98°C.

Helix: A spiral. Many large natural molecules (*e.g.* proteins and nucleic acids) are helical in shape.

Helmhotz Function (Helmhotz Free Energy): Symbol F. thermodynamic function defined by

$$F = U - TS$$

where U is the internal energy, T the thermodynamic temperature, and S the entropy. It is a measure of the ability of a system to do useful work in a feathernal process.

Helt Volard-Zelinky Reaction: A method for the preparation of halogenated carboxylic acids using free halogen in the presence of a phosphorus halide. The halogenation occurs at the carbon atom adjacent to the–CO OH group with Br_2 and PBr_3.

$$RCH_2COO = RCHBrCOOH$$
$$RCBe_2 = COOH$$

Hemihydrate: A crystalline compound with one molecule of water of crystallization per two molecules of compound, *(e.g.,* $2Ca\ SO_4.H_2O$)

Hemoglobin: The iron containing protein that is responsible for oxygen transport in the blood.

Henry Symbol H. The SI unit of inductance equal to the inductance of a coiled circuit that has a magnetic flux of one weber per ampere of current in the circuit.

$$1\ H = 1\ wb\ A^{-1}$$

Henry's Law: The solubility of a gas in a liquid is proportional to the pressure of the gas at a certain given temperature, mathematically it can be written as

Where C_A indicate the solubility of gas and C_B is its density which in turn in proportional to the pressure.

Heparin: A Complex organic acid related to the polysaccharides but containing sulphur and nitrogen which prevents the clotting of blood by interfering with the formation, and action of thormin. It is used as an anticoagulant.

Heptahydrate: A crystalline hydrated compound contains done molecule of compound per seven molecules of water of crystallization.

Heptaline ($C_{12}H_{10}$): Formed as an airsensitive, reddish brown, liquid which is readily polymerized by oxygen or by warming to 50°C.

Heptane (C_7H_{16}): The seventh member of the alkane series. It is found in petroleum and has nine known isomers; n-haptane has b.p. 98.4°C and r.d. 0.68.

Herbicides, Benzonic: A group of halo and halonitro derivatives of benzoic acid that control weeds, e.g. 2, 3, 6 trichlorobenzoic acid (TBA), 2,5 dichloro 3 nitrobenzoic acid (dinoben), 2, 3, 5 -triodo benzoic acid (TIBA), 3, 6 dichloro-2-meth-oxybenzonic acid (dicamba) etc. These are selective herbicides used for control of weeds in small grains, maize, sugarcane and turf. These are auxin type grcwth herbicides used in post emergence application against broad leaved annual and perennial weeds. Dicamba is active even against phenoxy acid resistant weeds.

Herbicides, Carbamate: Esters of N-aryl carbamic acid and aryl esters of N-alkyl carbamic acid which possess herbicidal activity, e.g. 2, 6-dit-butyl-4-methylphenyl N-methyl carbamate, (terbu-carb) used for the control of some grass weeds in cotton, maize, tomato and soybean. Other commonly used carbamate herbicides are dichlormate, chlorpropham, CEPC, proximpham, phenmedipham and desmedipham. Almost all the carbamate herbicides inhibit mitosis of plant cells followed by inhibition of photosynthesis. From the view point of environmental pollution, carbamate herbicides are non hazardous owing to their rapid degradation in plants and soils.

Herbicides, Dinitroaniline: A group of systemic herbicides absorbed by roots; the derivatives of N-alkyl or N, N -dialkyl- 2, 6-dinitro-aniline, Some important members of this group are : N-butyl N ethyl 4-trifluoro-methyl 1-2, 6 dinitroani line, benfluralin; 4-methylsul-phonyl-2, 6dinitro N,N-dipropylaniline, nitralin and N,N-dipropyl-4-trifluoro-methyl-2,6dinitroaniline, trifluralin. They reduce root growth, RNA, DNA levels and carbohydrate synthesis in plants; inhibit mitosis and thereby cause death of sensitive plants.

Herbicides, Halogenated Alkanoic Acid: A group of chlorinated aliphatic acids, e.g. trichloroacetic acid and 2, 2- dichloropropionic acid etc. used for weed control. Sodium salt of monochloro-acetic acid ($ClCH_2COONa$) is used for selective control of grass weeds in maize, potato and onion crops. The mode of action of this group of herbicides is the structural modification of proteins resulting in the phytotoxic action.

Herbicides, Nitrodiphenyl Ether: The nitro or chloronitro derivatives of diphenyl ether; used for weed control. For example, 2, 4-dichlorophenyl-4' nitro-phenyl ether (nitrofen); 4'-nitrophenyl 4-trifluromethyl -2-nitrophenyl ether (fluoro-difen) etc. are used for selective control of weeds in cereals. The preemergence application results in necrosis of shoot and meristems on sensitive weeds. After post-emergent treatment tilling begins within 24 hours and necrosis in 48 hours, resulting in death of sensitive plants thereafter.

Herbicides, Organophosphorus: Organophosphorus acids, esters or amides, used for weed control. For example O-(2, 4-dichlorophenyl) O-methyl N-isopropyl phosphoro-amidothioate (zytron) and O, O dibutyl 1-butylamino-cyclohcxyl phosphonate (aminophon) are used for preemergent control of grass weeds in legumes. N-phosphonomethyl glycine (glyphosate) is a wide spectrum nonselective postemergent herbicide with systemic action, These herbicides inhibit mitosis in plant cells.

Herbicides, Phenoxy Acids: A group of herbicides which are, derivatives of phenoxy alkanoic acids (including amides, esters and salts), e.g., 2, 4 dichlorophenoxyacetic acid (2, 4-D), 2, 4, 5 trichloro-phenoxyacetic acid (2, 4, 5-T), 2, 4-dichloroplie-noxy-propionic acid (2, 4 DP) etc., The salts of phenoxyacids are hormone type Systemic herbicides for postemergent control of dicot weeds in cereals, maize, rice and sugarcane. These herbicides affect almost all the biological processes of plants including protein synthesis resulting in abnormal growth and death of the plant.

Herbicides, Pyridazine: N-phenyl derivatives of 1, 2 pyridazine; used for preemergent control of broad leaved weeds and several grasses in cotton, beet and flax crops. For example, 5-amino-4-chloro -2-phenyl-3-pyridazone (pyrazon); 4 chloro-5-dimethylamino-2-(3-trifluoro-methylphenyl)-pyridazinone (metflurazone); 6-chloro 3-phenylpyridazin 4-yl-S-octyl thiocarbonate, (pyridate) and 3-(3-methyl-phenoxy) pyridazine (creda-zine) are some important herbicides of this group. These systemic chemicals are used at the rate of 2-4 kg active ingredient per hectare for control of grass weeds in rice, potato and strawberry and other crops. These compounds act by inhibiting the hill reaction in photosynthesis.

Herbicides, Pyridine: Substituted pyridines and picolinic acid derivatives, e.g., 2, 3, 5 trichloro 4-hydroxy pyridine (pyrichlor); 4-amino 3, 5, 6-trichloropicolinic acid (picloram); 3, 5, 6 trichloro, 2-pyridyloxy acetic acid (trichlopyr) etc. are used for selective weed control in cereals, maize, millet and Brassica cultures. They have systemic action like

Herbicides, Quaternary Ammonium Salts: Weedicides of compounds from N-alkyl hallides or sulphates of 2, 2 bipyridylium and 4, 4 bipyridylium and phenyl pyridinium ions, *e.g.,* paraquat, diquat and cyperquat.

Diquat is used for control of aquatic weeds. Cyperquat and paraquat are used for preemergent total weed control in soil. They are photosynthesis inhibitors and to some extent inhibit the growth of soil bacteria, but have little effect on other microbes. The herbicidal action of these compounds can be increased by nonionic surfactants.

Herbicides, s-Triazine: Derivatives of 1, 3, 5-s triazine form one of the most important groups of weedicides. Some well established members in the group are : 2-chloro 4,6-diethyl amino striazine (simazine); 2-chloro-4-ethylainino-6-isopropy-laminostriazine (atrazine); 2-chloro-4 , 6-di(isopropy-lamino)-s-triazine (pro-pazine); 2-chloro-4-ethyl-amino-6-diethylamino-s-triazine (trietazine); 2-chloro-4, 6, di (diethylamino-s-triazine (chlorazine) and 2-chloro-4, 6-bis(di-isopropy-lamino)-s-trazine Used for pre and postemergence weed control of dicot weeds in maize, sugarcane, tobacco, potato, millet and peanut. These herbicides kill weeds by inhibiting photosynthesis.

Herbicides, Sulphonyl Urea: These are derivatives of benzene sulphonic acid and urea, *e.g.,* I (2-chlorophenyl sulphonyl) -3-(4-methoxy-6-methyl-1, 3, 5-triazin-2-yl) urea (chlorsulfuron) and (2-methoxy-carboxypheny-lsulphonyl) 3-(4, 6-dimethyl-1, 3-pyrimidin-2-yl) urea, (sulfometuron). These are used to control broad leaved weeds in cereals. These herbicides inhibit the light dependent part of the photosynthesis.

Herbicides, Thiocarbamate: A group of herbicides with general formula $R_1S\text{-}CO\text{-}NR_2R_3$ where R_1 is alkyl, chloroalkyl or 4-chloro-benzy group and R_2 and R_3 may be alkyl or azepine groups, For example, EPTC cycloate, butylate, ethiolate, vernolate, diallate, benthio-carb and molinate are some of the important thiocarbamate herbicides. These are used for selective control of grass weeds in cereals-wheat and barley. Most of these inhibit protein synthesis and are mitotic poisons.

Herbicides, Uracil: Uracil derivatives substituted at 3, 5 and 6 positions, e.g., 3tert-butyl 5-chloro-6-methyl uracil (terbacil); 5-bromo-3-isopropyl-6-methyl uracil (isocil); and 3-sec-butyl 5-bromo-6-methyl uracil (bromacil); used for pre plant and pre emergent control of grass weeds in citrus, beet, spinach and strawberries. They inhibit the photosynthesis in weeds.

Herbicides, Urea: The N-alkyl,N, N'-dialkyl ureas and N-aryl N, N' dialky ureas form one of the most important groups of herbicides (along with striazines and phenoxyacids). Some of the important members of the group are 3-(4-hloro-phenyl)-l, 1-dimethyl urea (monuron); 1, 1-dimethyl 3-phenyl urea (fenuron); 3-(2, 4-dichlorophenyl) 1, 1-dimethyl urea (diuron); 3-(3-chloro-4-methoxyphenyl) 1, 1-dimethyl urea (metoxuron) and 3-(4-trifluorometh-lylphenyl)- 1, 1-dimethyl urea (parafluron). Used for both pre and post emergent control of grass and broad leaved weeds in Cotton, sugarcane and winter cereals, These act by inhibiting Hill reaction and Photosynthesis in plants.

Hess's Law: The change in enthalpy for any chemical reaction is constant whether the reaction occurs in one step or in several steps.

Hertzian Waves: Electromagnetic radiation covering a range of frequency from above 3 m 10^{10} hertz, corresponding to the shortest radar waves of I cm, to below 1.5 m 10 hertz, corresponding to long radio waves of 2000 metres.

Hertz: (Symbol $H_{2)}$ The SI unit of frequency, defined as one cycle per second (S^{-1}). Note that the hertz is used for regularly repeated processes, such as vibration or wave motion.

Heterocyclic Compound: A compound that has a ring containing more than one type of atom. Commonly, heterocylic compounds are organic compounds with at least one atom in the ring that is not a carbon atom. Pyridine and glucose one examples. The noncarbon atom is called a hetero atom.

Heterogeneous Catalyst: A substance that catalyzes a reaction occurring in another phase, *e.g.*, a solid substance might act as heterogeneous catalyst for a gas phase reaction.

Heterogeneous Reaction: A reaction which takes place between substances in different phases, *e.g.*, between a gas and a liquid.

Heterolytic Fission (Heterolysis): The breaking of a covalent bond so that both electrons of the bond remain with one fragment. A positive ion and a negative ion are produced

$$RX \rightarrow R^+ + X^-$$

Heterozygous: A condition in which the two factors (genes) determining a character or characters, are dissimilar; an organism may be heterozygous for one or several genes.

Heusler's Alloys: Alloys containing neither iron, nickel nor cobalt that exhibit strong ferromagnetism. They are composed of copper, manganese, and aluminium. Named after Conrad Heusler.

Hexadecane (Cetane $C_{16}H_{34}$): A colourless liquid alkane, m.p. 18°C, b.p. 287°C. Used as a solvent and in the determination of cetane numbers.

Hexadecanoic Acid or Palmitic Acid ($C_{18}H_{32}O_2$): A colourless waxy solid (m.p. 337°K). It occurs as glyceryl ester in palm oil. A solid mixture of palmitic and stearic acids, 'strearin' is used in making candles and soaps.

Hexafluropropene (C_3F_6): Obtained commercially by low pressure pyrolysis, of C_2F_4. Monomer used to form polymers.

Hexahelicene or Phenanthro (3, 4-C) Phenanthrene ($C_{28}H_{16}$): Yellow plates (m.p. 240-242°C). Steric crowding stops ring A and B becoming coplanar and they acquire a staggered configuration.

Hexaflurobenezene (C_6F_6): A colourless mobile liquid (b.p. 80°C) obtained C_6Cl_6 and KF in polar solvent.

Hexan-2, 5 dione or acetonyla-cetone ($CH_3COCH_2CH_2CO\ CH_3$): A colourless liquid (b.p. 191°C) obtained by boiling 2, 5-dimethylfuran with dilute sulphuric acid. It quickly condenses with a variety of substances to give derivatives of furan thiopen and pyrrole, and is solvent for cellulose acetate.

Hexane (C_6H_{14}): The sixth member of the alkane series. It is found in petroleum and has five known isomers; n-hexane has b.p. 69°C and r.d. 0.66.

Hexanoic Acid [Caproic Acid $CH_3(CH_2)_4COOH$]: A colourless oily liquid with an unpleasant smell, b.p. 205°C. It esters, e.g., ethyle caproate, are used in artificial flavourings.

Hexaploid: Having six sets of chromosomes (genomes); chromosome number of 6n.

Hexose: A monosaccharide whose molecule contains six carbon atoms, e.g.., the sugars, glucose, fructose and galactose.

Hexylresorcinol [$C_6H_{13}C_6H_3\ (OH)_2$]: A yellow crystalline solid m.p. 60°C, used as an antiseptic and in medicine.

High Pressure Liquid Chromatography: Also called high performance liquid chromatography (HPLC) is an instrumental liquid chromatographic technique for separation of less volatile polar compounds which cannot be separated by GC. In its simplest form it consists of a small chromatographic column containing adsorbent on which the sample is loaded. The sample is eluted with solvents under high pressure using a special pumping system.

HighSpeed Steel: A very hard steel containing 12%-22% tungsten, with chromium, vanadium, molybdenum and small amounts of other elements; used for tools which remain hard even at red heat.

High Spin State: A term used to indicate the maximum number of electrons of the compound with the electronic configuration and stereochemistry. Most commonly used for octahedral complexes where depending upon the crystal field splitting two spin states are possible for.

Hippuric Acid or N-Benzoylglycine ($C_9H_9NO_3$): Solid with melting point 187°C, soluble in water and alcohol. It is excreted in small quantity in the urine of mammals.

Histamine [$C_3H_3N_2(CH_2)_2NH_2$]: A white crystalline substance (m.p. 86°). It occurs in animal tissues.

Histidine: A crystalline soluble amino acid which occurs in fish and from which histamine is manufactured.

Hoesch Synthesis: A reaction in which condensation of polyhydric phenol takes place with nitrites by the action of hydrochloric acid (with or without $ZnCl_2$ as catalyst). This results an imine hydrochloride which on hydrolysis with water results the hydroxy-ketone.

Hofmann Degradation: The reaction in which amide changes into amine in the presence of aqueous KOH and bromine

$$RCONH2 \xrightarrow{Br2/KOH} RNH_2$$

Holnium: A soft maleable silvery element of the lanthonoid series of metals. It occurs in association with other lanthanoids. It has few applications.

Symbol HO; m.p. 1474°C; b.p. 2695°C; n.d. 8.80; p.n. 67; r.a.m. 164.93.

Homeomorphous: Having the same crystalline form but different chemical composition.

Homocyclic Compound: A compound containing a ring made up of the same atoms. Benzene is an example of homocyclic compound.

Homogeneous: Relating to a single phase. A homogeneous mixture, for instance, consists of only one phase.

Homogentistic Acid ($C_8H_8O_4$): Excreted in urine in the rare hereditary disease alkaptonuria. It is easily oxidized in air to darkcoloured polymeric products.

Homologous pair: In spectrographic analysis a homologous pair consists of the particular spectral line utilized in the determination of the concentration of the element and an internal standard line, such that the ratio of the intensities of the radiation producing the lines remains unchanged with the variation in the conditions of excitation.

Homologous Series: A group or class of organic compounds each containing one particular characteristic group (functional group. Some characteristics of the series are that they have general formula, common difference of CH2, general methods preparation and have similar chemical properties, *e.g.*, homologous series of alcohols.

$C_n H_{2n+1} OH$	General formula
$CH_3 OH$	Methyl alcohol
$C_2 H_5 OH$	Ethyl alcohol
$C_3 H_7OH$	Propyl alcohol

Homologues: Members of the same homologous series, *e.g.* ethyne, C_2H_2 and propyne C_3H_4.

Homolysis: The breaking of a covalent bond into two free radicals such

that R:R → R. + R.

Homolytic Fission: The symmetrical breaking of a covalent bond so that one electron is acquired by each atom, *e.g.*, dissociation CI_2 molecule in the presence of sunlight Cl-Cl → 2Cl (free radical).

Hormones: Compounds secreted directly into blood stream by certain organs of the body, *e.g.* the sex hormones secreted by testis and ovary and insulin which is secreted by certain parts of the pancreas.

Hot Working: Deformation of metal and alloys carried out above the recrystallization temperature.

Houben-Hoesch Synthesis: By condensation of cyanides with polyhydric phenols (m-isomers particularly) in presence of $ZnCl_2$ and HCl.

HPLC: High-Performance liquid chromatography. This is a sensitive analytical technique similar to gas liquid chromatography, but using a liquid carrier which is chosen specifically for a particular substance to be detected.

Humectant: A compound which due to its hydroscopic nature can be used to preserve the moisture content of materials, e,g., glycerol or sorbitol in confectionery or food.

Hume-Rothery's Rule: The statement that the phase of many alloys is determined by the ratios of total valency electrons to the number of atoms in the emperical formula.

Humic: Pertaining to the soil or water borne substances resulting from the partial decay of leaves, plant material or any other organic waste.

Humic Acid: A group of aromatic acids of high molecular mass and unknown structure that are present in solid and can also be formed from peat and brown coal or after refining sugar beet.

Humidity: The humidity of moist air may be expressed several ways: The relative humidity is the ratio of the partial pressure of water vapour in the air to the partial pressure of water vapour in the air when saturated at the same temperature. The absolute humidity is the mass of water vapour per unit mass of dry air.

Humus: A dark coloured amorphous material obtained by the microbiological decomposition of plants material, mainly lignin and proteins. It occurs in soil in company with clay.

Hund's Rule: The electrons are distributed among the orbitals of a subshell in a way that gives the maximum number of unpaired electrons with parallel spins.

Hybridisation: The phenomenon of rearrangements and combination of an atom to give new identical orbitals in a molecule. The new orbitals are called hybrid orbitals.

Hydrazine or Diamide (NH_2NH_2): A colourless liquid obtained by treating a solution of potassium sulphate saturated with nitric oxide with sodium amalgam.

$$K_2SO_4 + 2NO \rightarrow K_2SO_3 + N2O_2$$

$$K_2SO_3.N_2O_2 + 6H \rightarrow K_2SO_4 + N_2H_4 + H_2O$$

(From Na/Hg and water)

It is used as a fuel for rockets and reducing agent,

Hydrazobenzene or 1, 2-Diphenyl Hydrazine (C12H12N2): A colourless plates (m.p. 131^0 C obtained by reduction nitrobenzene with iron and sodium hydroxide. It is insoluble in water, easily soluble in organic solvents. It is widely used for the preparation of benzidine.

Hydrazoic Acid (N_3H): A colourless volatile liquid obtained by the action of nitrous acid on hydrazine or its salt, e.g., hydrazine chloride.

$NH_2NH_2 + HNO_2 \rightarrow N_3H + 2H_2O$ It is poisonous in nature.

Hydrazone: Organic compounds formed by the reaction of hydrazine with aldehydes or ketones in presence of an acid. They have general formula $RCH = NNH_2$ or $RR'C = NNH_2$, where R and R are alkyl or aryl groups.

Hydriodic Acid or Hydrogen Iodide (HI): A colourless; pungent smelling gas obtained when hydrogen and iodine mixture is heated in a tube containing finely divided platinum as catalist.

$$H_2 + O_2 = 2HI$$

It is used as a reducing agent in organic chemistry.

Hydrobromic Acid (HBr): A colourless pungent smelling gas obtained by the hydrolysis of phosphorus bromide, It is highly soluble in water. It is used as laboratory reagent in organic chemistry.

Hydrochloric Acid (HCl): A colourless gas obtained by heating sodium chloride with concentrated sulphuric acid. It has a sharp pungent odour and is very soluble gas. It reacts with bases to form salts. It is used in the manufacture of chlorine, chlorides and glucose from corn starch. It is used in the preparation of aqua regia and medicines. It is also used as laboratory reagent.

Hydrocracking: A process which involve both the hydrogenations and catalystic cracking of petroleum fractions, resulting high quality gasolines.

Hydrocyanic Acid or Hydrogen Cyanide (HCN): A colourless colatile liquid (b.p. 299 K) obtained by passing a mixture of Carbon monoxide and ammonia over heated alumina (Catalyst) at 770-970 K.

$$CO + NH_3 \rightarrow HCN + H_2O.$$

It smells like bitter almonds and is deadly poison. It is used in organic synthesis and also used in killing insects and rats. It is believed to exist in two tautomeric forms as given below :

Hydrofluoric Acid (HF): A colourless liquid obtained by heating fluorspar (calcium fluoride) with concentrated sulphuric acid. It is highly poisonous and very corrosive to skin. It is a weak acid, weaker than phosphoric acid. It is used for removing silica, as antiseptic in brewing industries, in etching of glass and fluorides are used as an insecticide.

Hydrogen: A chemical element belonging to group I. It is a colourless gas obtained in the laboratory by the action of dilute HCl or H_2SO_4 on zinc metal. It is a strong reducing agent and combines with nonmetals. It is used in the conversion of liquid vegetable oil to a solid fat in the presence of nickel. It is also used to convert coal into a liquid fuel and lubricant by hydrogenation. It reacts with metal oxide to convert them into metals. In addition to this, hydrogen is used in the synthesis of ammonia, methanol and formaldehyde.

Symbol : 14 ; a.n. 1 ; m.p. 259°C ; b.p. 252°C ; d, 0.0899 kg m^{-1}.

Hydrazine (N_2H_4): A colourless liquid that can be prepared by the oxidation of ammonia with sodium chlorate 10 or by the gas phase reaction of ammonia with chlorine. Hydrazine is a weak base forming with strong acids, e.g.., N_2H_4HCl.

Hydride: A compound of hydrogen. Ionic hydrides are formed with highly electropositive elements and contain the H ion, Nonmetals form covalent hydrides as in methane or silane.

Hydrocarbon: Any compound containing only the elements carbons, and hydrogen. Examples are alkanes, alkenes, alkynes and aromatics such as benzene and naphthalene,

Hydrogenation: Subjecting to the chemical action of, or· causing to combine with, hydrogen.

Hydrogen Bond: An inter molecular or intra molecular bonding in which hydrogen is bonded to a strongly electronegative element, such that this element polarizes the bond, producing a net positive charge on hydrogen which gets temporarily attached to another electron rich centre in the same molecule or of some other molecule,

Hydrogenation of Coal: The manufacture of artificial mineral oil from coal by the action of hydrogen. This involves causing the carbon in coal to combine with hydrogen to form hydrocarbons.

Hydrogenation of Oils: Artificial hardening of liquid animal and vegetable Oils by the action of hydrogen. Liquid fats and oils contain a high percentage of liquid triolein, $C_{57}H_{104}O_6$, which may be converted into a solid tristearin, $C_{57}H_{106}O_6$, by the action of hydrogen in the presence of a finely divided catalyst; the result being a hard fat of higher melting point.

Hydrogen Bonding: In the case of water, the polarity give rise to an intermolecular force whose power is unique. The oxygen atom in a water molecule is not only bonded to two hydrogen atoms, but it is attracted to hydrogen atoms belonging to other molecule as well. It is a very strong intermolecular force present in compounds containing -OH and -NH group as well as in hydrogen fluoride, H.F.

Hydrogen Electrode: A primary standard electrode which consists of small platinum strips Pure hydrogen at coated with platinum black sealed through a glass tube. The atmospheric pre electrode is placed in a dilute solution of an acid and pure hydrogen gas is then admitted into it at one atmosphere pressure.

Hydrogen Ion Concentration: The number of grams of hydrogen ions per litre of solution. It is useful as a measure of the acidity of a solution and in this context is usually expressed in terms of pH = $\log_{10} 1/[H^+]$ where [H] is the hydrogen ion concentration As pure water at ordinary temperatures dissociates slightly into hydrogen ions and hydroxyl ions (H_2O = H + OH), the concentration of each type of ion being 10^{-7} mole per litre, the pH of pure water will be logiol/10^{-7} = 7. If acid is added to water, its hydrogen ion concentration will increase and its pH will therefore decrease. Thus a pH below 7 indicates acidity and similarly a pH in excess of 7 indicates alkalinity.

Hydrogenolysis: The cleaving of a chemical bond by hydrogen, generally in the presence of a hydrogenation catalyst.

Hydrogen Peroxide: A colourless liquid formed by the action of dilute sulphuric acid on sodium peroxide or barium peroxide. It is miscible with water, alcohol and ether in all proportions. It acts as an oxidising and reducing agent. It is a used in bleaching delicate things like wool, silk, ivory and feathers. It is also used as antiseptic, germicide, as a propellant in rockets, as an oxidizing agent in the laboratory and washing wounds, teeth and ears.

Hydrogen Sulphide (H_2S): A colourless gas obtained in the laboratory by the action of hydrochloric acid on iron (II) sulphide. Its odour is like that of rotten eggs. It is weak acid and reacts with many salt solutions to give coloured sulphides. It is used in qualitative analysis.

Hydrogen of Salt: A reaction in which the anion or cation of salt react with the solvent to produce acidity or alkalinity, e.g., hydrolysis of sodium carbonate

$$Na_2CO_3 + 2H_2O \rightarrow 2NaOH + H_2CO_3$$

Hydrophobic Bonding: The association of nonpolar molecules or groups' in aqueous media, resulting from the tendency of water molecules to exclude nonpolar species.

I

Ice: Water, H_2O in the solid state, formed at the freezing point of water 0°C. As it is less dense water, water expands on freezing and ice floats on water.

Ice Point: The temperature of equilibrium between ice and water under normal atmospheric pressure; i.e., the melting point of ice. The ice point is assigned the value of 0°C.

Icosahedral: A solid figure with twelve apices, each face triangular. It is found in many derivatives, e.g., $B_{12}H_{12}^{2-}$

Ide: Suffix denoting the chemical name or the anion radical of the binary compound formed by atoms of two elements only, e.g., hydrogen sulphide, H_2S, calcium carbide, CaC_2, Sodium chloride, NaCl etc.

Ideal Crystal: A crystal whose lattice is perfectly regular and contains no foreign atoms or ions or other defects or imperfections.

Ideal Gas: A theoretical concept of gas which would obey the gas laws exactly. Such a gas would consist of perfectly elastic particles.

Ideal Solution: A solution obeying Raoult's Law. Such a solution would be obtained from its components with zero heat of mixing (DH=O), zero volume change (DV=O) and ideal entropy of mixing. Igneous Rock : A rock which has solidified from liquid magma.

Ignition: The act of setting fire to something. Initiating combustion by raising the temperature of the reactants to the ignition temperature; particularly the process or means of firing the explosive mixture in an internal-combustion engine by an electric spark.

Ignition Temperature or Point: The temperature to which a substance must be heated before combustion can take place.

Ilkovic Equation: The relation between diffusion current id and the concentration c in photography which is its simplest form is

$$id = k'c$$

Ilmenite ($FeTiO_3$): A back *Ti* ore occurring in sands.

Imide Group: Compounds containing—CONHCO— group (an imino group between two carbonyls), e.g. succinimide. These may be formed by reaction of an amine with one molecule of a dicarboxylic acid or two molecules of monocarboxylic acids (at. imido group)

Imine: A compound containing the group—NH-, where the nitrogen atom is not joined to carbonyl groups or to other hydrogen atoms.

Immiscible: Incapable of being mixed to form a homogeneous substance; it is usually applied to liquids; e.g., oil and water are immiscible.

Indicator: A coloured substance usually derived from plant materials that can exist in either an acid or base form. The two forms are differently coloured. By adding a small amount of at', indicator to a solution and noting its colour, it is possible to find out whether it is in the acid or base form. Common examples of indicators are Methyl violet, Methyl orange, Phenolphthalein etc.

Indigo or Indigotin ($C_{16}H_{10}O_2N_2$): A dark blue powder (m.p. 663-665 K), obtained by condensation of aniline with chloroacetic acid to get N-phenylglycine which is fused with caustic sodamide at 570-670 K to produce indoxyl. This on oxidation gives indigo.

Indium: A soft silvery metallic element belonging to group III of the periodic table. It is found in minute quantities. Primarily in zinc ores and is used in alloys, in several electronic devices, and in electroplating.

Symbol : In; m.p. 156.61°C; b.p. 2080°C; r.d. 7.31; p.n. 49; r.a.m. 114.82.

Indole-3-Acetic Acid ($C_{10}H_9NO_2$): A white crystalline substance, m.p. 168-170°C, that promotes plant growth.

Indole or Benzopyrrole (C_8H_7N): A colourless crystalline solid (m.p. 325K) obtained by heating o-amino-chlorostyrene with sodium ethoxide.

Indoxyl: A bright yellow solid (m.p. 358 K) obtained by hydrolysis of indican. It is readily by air to indigotin.

Inductive Effect: The electron displacement along a carbon chain due to the presence of electron repelling or electron withdrawing group.

Industrial Methylated Spirits (IMS): Ethanol rendered nonpotable by mixing 95% alcohol with wood naphthas.

Inert Gas: A semiobsolete name for the zero group elements.

Inert Gases: The six chemically inactive gaseous elements belonging to group 18-He, Ne, Ar, Kr, Xe and Ra.

Inert Pair Effect: An apparent correlation between the existence of oxidation states m and (m-2) and the s^2 configuration for all latter for the heavy p block elements, e.g., Pb shows oxidation state Pb (IV) and Pb (11).

Infra-Red Radiation: Radiation of frequency less than about 10^{13} per sec, i.e., of wavelengths greater than about 8000A. It is not visible to human eye.

Infusible: Difficult to melt; having a very high melting point.

Infrared (IR) Spectroscopy: The various stretching and bending vibrations of a valency bond, between two atoms, occur at certain quantized frequencies. When infrared light of the same frequency is incident oil the molecule, energy is absorbed and amplitude of that vibration increases. The study of this absorption spectrum gives an important clue about the nature of bonds and the functional groups in the molecule. Very small amounts (one milligram or less) of a pure sample as a solution, in a sodium chloride cell, or as a thin film on sodium chloride plate, in a KBr pellet is scanned by infrared radiation of increasing wavelength or decreasing frequency (4000-400 cm), and a spectrum is recorded. The spectrum, which displays absorption maxima at various wavenumbers, e.g. for C—H in alkane, 2850-2960 and In alkene 3000-3100; C—C in alkene, 1620-1680, aromatic, about 1600; carbonyl stretching, 1705-1725; unsaturated aldehyde carbonyl, 1660-1680, O—H of alcohol, 3590-3650; NH in amine 3300-3500 and CN stretching at 2200-2240 cm^{-1}. The same bonds show vibration absorptions at still lower frequencies of lesser intensity. The fingerprint region (1430-910 cm^{-1}) in IR spectrum of any compound is generally not reproducible by other compounds.

Inhibitor: A substance that decreases the efficiency of a catalyst.

Inks: Deeply coloured liquids of varied composition; many black ar, blue-black inks own their colour to organic compounds of iron.

Inorganic: Pertaining to matter that is neither living nor immediately derived from living matter. All chemicals derived from metals and nonmetals are inorganic except the compounds of carbon. But ionic compounds of carbon including those devoid of hydrogen (e.g. CO and CO_2) are included in this group.

Inorganic Chemistry: The branch of chemistry, which deals with the study of metallic and nonmetallic elements and their compounds.

Insecticide: A substance used for killing insect pests.

Insecticides, isobutylamides: A group of naturally occurring carboxylic acid amides of isobutylamine with insecticidal action; for example pellitorine, fagaramide, anacycline, spilanthol, herculine and piperlongumine etc. They are known for their high knockdown effect and unsatisfactory stability in the environment.

Insecticides, Organo Carbamate: A group of broad spectrum insecticides derived from aryl or alkyl N-methyl carbamates. They are known choline esterase inhibitors and neurotoxins. Some important members of this group are : 1-naphthyl N-methyl carbamate, (carbaryl); 2-isopropylphenyl N-methyl carbamate (MIPC); 2, 3, dihydro-2, 2-dimethyl benzofuran 7-yl N-methyl carbamate and 2-methyl-2-methylthiopropanaldehyde O-methyl carbamoyl oxime.

Insecticides, Organochlorine: chlorinated hydrocarbons mostly derived from benzene, or hexachlorocy-clopentadiene; with a broad spectrum of insecticidal activity. But for lindane or y-hexachlorocyclohexane, the rest of them can be classified into two major groups, e.g., DDT analogues and cyclodiene insecticides. Of these DDT has been the most widely used insecticide, the worldover, although its analogues, viz. DFDT, prolan, bulan etc. have not proved promising. Hexachlorocyclohexane is the maximum produced commercial insecticide to date, although only gisomer (lindane) of this compound has shown to be insecticidally active. At present DDT and the cyclodiene insecticides like aldrin, endrin, dieldrin have lost popularity due to their persistence in the environment and property of accumulation in animal fat, and hence are facing ban in a number of countries. .

Insecticides, Organophosphorus: Alkyl and aryl esters and amides of phosphoric, and thiophosphoric acids which act as nerve poisons and anti insect agents. As a matter of fact, major compounds of this group do not contain a PC bond and hence are not organo phosphorus compounds in true sense. Therefore, these are referred to as organophosphate insecticides (thiophosphates and dithiophosphates or their amides are also included). Important compounds of this group are: tetraethyl pyrophosphate (TEPP); hexaethyl tetraphosphate (HETP); octamethyl pyrophosphoroamidate (OMPA); O, O-diethyl O-4-nitrophenyl phosphorothioate (parathion); O, O-diethyl O-(2-ethyl thioethyl) phosphorothioate (demetonO); O, O-dimethyl-O-(2, 2-dichlorovinyl) phosphate (dichlorvos, DDVP); O, O-dimethyl-S-(I, 2-dicarbethoxy) ethyl phosphorodithioate (malathion) and O-ethyl-S-phenyl ethylphosphono dithioate (fonofos). Organophosphorus insecticides are known to be cholinesterase inhibitors.

Insoluble: Describing a compound that has a very low solubility (in a specified solvent).

Intermediate Bonding: A form of covalent bond that also has an ionic or electrovalent character.

Intermolecular Forces: Forces of attraction between molecules rather than forces within the molecule (chemical bonding). If these inter molecular forces are weak the material will be gaseous and as their strength progressively increases materials become progressively liquids and solids.

Internal Energy: Thermodynamic energy U. The total energy associated, with a system, which cannot itself usually be determined. However, the change in the internal energy of a system, DU, is useful, thermodynamic quantity, and in defined by DU = $q - w$, where q is the heat abstracted by the system from its surroundings and w is the work done simultaneously on the surroundings.

Internal Resistance: Resistance of a source of electricity. In the case of a cell, when a current is supplied the potential difference between the determinals is lower than the e.m.f. The difference is proportional to the current supplied. The internal resistance (r) is given by

$$r = (E - V)/I$$

where E is the e.m.f., V the potential difference between the terminals, and I the current.

Interstice: A small space between the ions, atoms or molecules in a crystal lattice, e.g., in steel, there are atoms of carbon in the interstices of the iron atoms forming the lattice of the metal crystals.

Interstitial Compound: A crystalline compound in which atoms of a nonmetal (e.g., carbon, hydrogen, or boron) occupy in terstitial positions in the crystal lattice of a metal. Interstitial compounds are often nonstoichiometric they have metallic lustre and are electrical conductors.

Inversion: A change from one optical isomer to the other.

Iodic (V) Acid (Iodic Acid, $H1O_3$): A colourless crystal obtained by the oxidation of iodine with fuming nitric acid or treating a suspension of iodine with chlorine.

It is soluble in water and insoluble in alcohol. On heating, it decomposes to give iodine pentaoxide. It is an oxidizing agent and oxidizes hydriodic acid to iodine. It gives a, number of salts, e.g., KIO_3.

Iodic (VII) Acid (Periodic Acid, $H1O_4.2H_2O$): The common periodic acid is paraperiodic acid, $HIO_4.2H_2O$ or H_5IO_6. It is obtained by heating a suspension of iodine in per chloric acid.

A colourless crystalline, deliquescent solid which is readily soluble in water and slightly soluble in water and ether. It is a monobasic fairly strong acid. Iodides are reduced to iodate or iodine. It decomposes into iodine penta oxide with evolution of water and oxygen.

Iodine: A lustrous black solid belonging to group 17, s.p.g. 4.94. It is sparingly soluble in water but dissolves in concentrated solution of potassium iodide.

It is used in the preparation of sodium tetrathionate, manufacture of dyes, drugs, iodoforms and also used in the laboratory to estimate thiosulphate.

symbol: 1; a.n. 53; m.p. 113.5°C; b.p. 184°C; r,d, 9.95.

Iodine Dioxide or Tetraoxide ($1O_2$ or $12O_2$): A lemon yellow solid obtained by the action of hot concentrated sulphuric acid over iodic acid or by oxidizing iodine with cold nitric acid.

$$H_2SO_2\ 4HIO_3 \rightarrow 2H_2O + I_2O_4 + 3O_2.$$

It is slightly soluble in cold water. With hot water, it forrns a solution of iodic acid with liberation of iodine.

$$I_2O_4 + 4H_2O \rightarrow 8HIO_3 + 1_2$$

Iodine Monochloride: A dark red liquid made by passing chlorine radioactive. It has properties similar to those of its constituents halogens. Iodine monochloride used as a nonaqueous solvent and as an iodating agent in organic reactions.

Iodine Number or Iodine Value: A measure of the degree of unsaturation of a product, such as, an oil or fats. It is expressed in grams of iodine absorbed by 100 g of the given substance.

Iodine (V) Oxide (Iodine Penta Oxide, 12Os): A white crystalline solid. It is fairly soluble in water to form iodic acid. It is a strong oxidising agent.

Iodine Pentanuoride (IF_5): A colourless liquid obtained by the direct combination of the elements or by heating together iodine and silver fluoride or by the interaction of fluorine and heated iodine pentaoxide.

Iodine Trichloride (IC_{13}): A lemon yellow crystalline solid (m.p. 374 K, under 16 atm. pressure) obtained by the action of excess of chlorine on iodine or the monochloride.

$$ICI + Cl_2 = ICI_3$$

It is soluble in organic solvents, liquid ammonia and dioxide. It forms crystalline additive compounds, with metallic chlorides of the type $MICl_4$ which dissociate on exposure to air.

Iodobenzene or Phenyl Iodide (C_6H_5I): A colourless, heavy liquid (b.p. 461 K). It is obtained by healing benzene and iodine with an oxidising agent (nitric or iodic acid).

Iodoethane (Ethyl Iodide, C_2H_5I): A colourless liquid alkyl halide made by reaction of ethane with iodine in the presence of the phosphorus.

Iodomethane: A liquid alkyl halide made by reaction of methanol with iodine in the depressive offered phosphorus.

Ion: An electrically charged atom or group of atoms. Positively charged ions are cations and negatively charged ions are anions.

Ionic Crystal: A crystal composed of ions of two or more elements. The positive and negative ions are arranged in definite patterns and are held together by electrostatic attraction.

Ionic Strength: The ionic strength of electrolyte solution is given

$$I = ½½\Sigma\ i\ mi\ zi^2$$

where z is the charge on the ionic species and m is the molar concentration, the summation being continued over all the different ionic species in the solution. Ionization : The process of producing ions. Most of the ionic salts give ions as soon as they are dissolved in water. The negatively charged ion is called anion and positively charged ion, is called cation. The driving force for ionization in solutions is salvation of the ions by solvent molecules, e.g. dissociation of HCI in water produces hydronium ions and chloride ions :

$$HCI + H_2O\ 4\ H_3O^+ + Cl^-$$

Ions can also be produced by ionizing radiation, i.e. the impact of a particle or photon can knock out an electron from an atom or a molecule to give cation ($A \rightarrow A^+ + e^-$); or an atom or molecule can capture an electron to form an anion ($A + e^- \rightarrow A^-$).

Ionisation Energy (IE): The energy required to remove the most loosely bound electron from an isolated atom of an element in the gaseous state. The factors influencing the ionisation energy are size of the atom, charge on the nucleus and screening effect. It is measured in electron volts.

Ion Product For Water: The product of the concentration of H+ and OH. So, the ionic product of water is

$$Kw = I \times 10^{-14} = [H^+]\ [OH^-]$$

where *Kw* is the ion-product constant for water.

Iridium: A silvery white metal belonging to group 9. It is brittle and fairly hard. It is used to harden platinum for, electrical contacts and in jewellery.

Symbol : Ir; a.n. 77; m.p. 2450°C; b.p. 4130°C; r.d. 22.4.

Iron: A transition greyish metal belonging to group 8. It occurs in nature as Haematite, Fe_3O4 ; Limonite, $Fe_2O_3.H_2O$: Magnetite, Fe_3O_4; iron pyrite, FeS_2; spathic iron ore, $FeCO_3$. It is extracted in blast furnace using coke, lime stone and hot air. It is very reactive toward air, nonmetals and acids.

Symbol : Fe ; a.n. 26 ; m.p. 1539°C ; b.p. 2887°C; r.d. 79.

Iron (II) Chloride (Ferrous Chloride, $FeCI_2$): A white crystalline anhydrous solid obtained on heating iron filings in a stream of dry hydrogen chloride, It is soluble in water as well as in methyl alcohol. The hydraled $FeC12.4H_2O$ is prepared by dissolving iron in dilute hydrochloric acid.

Iron (111) Chloride (Ferric Chloride, $FeCI_3$): A compound obtained in the anhydrous form as dark red crystals by passing stream of dry chlorine of heated iron filings. Its vapour density at 444°C corresponds to the formula Fe_2Cl_6 and at 750° to $FeCl_3$. On heating above 750°C; it breaks up into iron (11) chloride and chlorine, It is soluble in water, alcohol and ether. It is used in medicine and as laboratory reagent.

Iron (II) Sulphate (Ferrous Sulphate or Green Vitrol, $FeSO_4.7H_2O$): A green monoclinic crystal obtained commercially the action of air and water on iron pyrites at ordinary temperature

$$2FeS_2 + 2H_2O + 7O_2 \rightarrow 2FeSO_4 + 2H_2SO_4$$

It is isomorphous with epsom salt, $MgSO_4.7H_2O$, as well as with zinc sulphate, $ZnSO_4.7H_2O$. The anhydrous salt is white. It is an efficient reducing agent. It is used as mordant in dyeing, as an insecticide, in the preparation of ink and as laboratory reagent in volumetric analysis.

Iron (III) Sulphate [Ferric Sulphate, $Fe_2(SO_4)_3$]: A substance obtained in the anhydrous from by heating iron (11) sulphate with about fifteen times its weight of concentrated sulphuric acid. It forms double salt with sulphates of alkali metals, e.g., iron alum $(NH_4)2SO_4$ $Fe_2(SO_4)3.24H_2O$ which is used in volumetric analysis.

Iron (11) Sulphide (Ferrous Sulphide, FeS): A precipitate obtained by passing hydrogen sulphide, through an alkaline solution of a ferrous salt. The large scale preparation is done by heating iron filings with sulphur

$$Fe + S \rightarrow FeS$$

It reacts with dilute acids to give hydrogen sulphide and is used in the laboratory for the preparation of this gas.

Irreversible Reaction: A reaction in which conversion to products is complete. There is little door no back reaction.

Isatin: A red crystalline solid (m.p. 473 K) obtained by oxidation of indigotin with nitric acid.

Isinglass: A product containing about 90% gelatin, made from the swimming bladders of fish. Used for clarifying alcoholic beverages.

Iso-I: Prefix denoting an isomer with a branched chain.

Isobars: The nuclides that have the same mass number but different atomic number, e.g., ${}_{16}^{36}S$ and ${}^{36}_{18}Ar$.

Isobutyric Acid [$(CH_3)_2CHCOOH$)]: A colourless oily liquid (b.p. 427 K) obtained industrially by oxidation of isobutyl alcohol, It is easily oxidised to acetone or acetic acid and carbon dioxide.

Isocyanide Test: (Carbyl Amine Test) The primary amine is warmed with chloroform and alcoholic solution of potassium hydroxide, Isocyanide is formed which gives a very offensive smell

$$CHC1_3 + 3KOH + RNH_2 \rightarrow RNC + 3KCI + 3H_2O$$

Isoelectronic: The compounds having the, same number of electrons, e.g., carbon monoxide (CO) and nitrogen (N_2) are isoelectronic.

Isomers: A chemical compound having different chemical and physical properties but having the same molecular formula,

Isomerism: The phenomenon shown by two or more compounds with the same molecular formula but different physical and chemical properties.

Isomerism, Chain (Nuclear isomerism): A type of structural isomerism where isomers differ in the arrangement of carbon chain (i.e. normal, iso-or neo-skeleton) or having a primary, secondary or tertiary carbon, skeleton as in case of isomers of pentane:

Isomerism, Geometrical: (*cis*-trans or E, Z-isomerism) A type of stereoisomerism due to spatial arrangement of atoms or groups on the carbon atoms bearing a double bond, The isomer in which the similar groups lie on same side of the double bond is called the cis-isomer or Z-isomer and the other in which the similar groups lie on the opposite sides is called the trans or the E-isomer. Therefore, geometrical isomerism is called cistrans isomerism and E-Z isomerism, e.g., cis (fumaric acid) and *trans* forms of but-2-ene-1, 4-dicarboxylic acid.

Isomerism, Metameric: (Metamerism) Structural isomerism in a homologous series due to the difference in the nature of alkyl or aryl group attached to the polyvalent atom as in case of diethyl ether ($C_2H_5OC_2H_5$) and methyl propyl ether ($CH_3O\ C_3H_7$).

Isomerism, Optical: The isomerism shown by steroisomers which differ in the spatial arrangement of four different groups or atoms around a carbon atom (called *chiral* or asymmetric carbon or optical centre). The two optical isomers thus formed are non superimposable mirror images of each other (the *enantiomers*). A mixture of equal parts of enantiomers is called a *racemic modification*. If a compound has n optical centres or chiral carbons, it will possess 2^n optical isomers, The sterioisomers that are not mirror images of each other are called diastercomers. For example, glyceraldchyde having only one asymmetric carbon gives rise to two enantiomers only but glucose with 4 chiral centres gives rise to 16 optical isomers.

Isomerism, Position: Isomerism caused by difference in position of the same substituent (or functional group) in the same chain, e.g. in two forms of propanol.

$$CH_3CH_2CH_2OH \quad \underset{\displaystyle OH}{\underset{|}{CH_3CHCH_3}}$$

2-propaanol

isomerism, Tautomeric: A kind of structural isomerism due to the shift of a labile hydrogen between two polyvalent atoms, carbon, nitrogen, oxygen etc., e.g. isomerism in two forms of ethylacetoacetate. (at. tautomerism).

Isomorphism: The existence of compounds with the same crystal structure.

Isonitrile: An organic compound of the formula R—NC.

Isoprene [CH_2:CH.C(CH_3): CH_2]: A colourless liquid, b.p. 34°C. Natural rubber consists mainly of a polymer of isoprene.

Isosterism: Property of substances having molecules with the same number of atoms and the same number of electrons; indicating a similarity in physical properties, e.g. carbon dioxide, CO_2 and nitrous oxide, N_2O)

Isoquinoline (C_9H_7N): A colourless soid (m.p. 296 K) having an odour resembling benzaldehyde obtained by heating oxime of Cinnamaldehyde with Phosphorus pentaoxide.

Isostructural: Having the same lattice type and crystal structure, e.g., K_2PtCl_6 and Rb_2PtCl_6.

Isotherm: A line on a chart or graph joining points of equal temperature.

Isothermal Change: A process which takes place at a constant temperature.

Isotones: The nuclides with the same number of neutrons but different atomic number, e.g., $^{14}_{6}C$, $^{15}_{7}N$ and $^{16}_{8}O$.

Isotonic: Describing solutions that have the same osmotic pressure.

Isotopes: Atoms of a given element that differ in number of neutrons but having different masses, e.g., $^{14}_{6}C$ $^{15}_{7}N$ and $^{16}_{8}O$.

Isotopic Mass: The mass number of a given isotope of an element.

Isotopic Number: The difference between the number of neutrons in an atom and the number of proton.

J

Javelle, Water (Eau de Javelle): A solution containing potassium hypochlorite, KOCI; made by the action of chlorine on a cold solution of potassium hydroxide, KOH. Used for bleaching and as a disinfectant.

Jewellers Rough: Iron (111) oxide (Fe_2O_3), used as a mild abrasive.

John Reductor: A tube containing zinc amalgam. It is used for reduction of solutions (e.g., Fe^{3+} ® Fe^{2+}) prior to titration or estimation.

John-Teller Effect: The John-Teller theorem says that when any degenerate electronic state contains a number of electrons such that the degenerate orbitals are not fully filled, the geometry of the species will change so as to form non degenerate orbitals. It is generally applied to transition metal compounds where the d^9 state is d^6 d^6 e. Cu (II), generally has the distorted octahedral coordination typical of John-Teller effect.

Joule (Symbol J): The Sl unit of energy and work.

Joule's Law: The internal energy of a gas depends only on its temperature (being independent of its pressure and volume). At high pressure it is invalidated by the existence of intermolecular forces.

Joule-Thomson Effect: The change in temperature produced when a gas expands by flowing through some kind of obstruction (like a porous plug) from a region of higher pressure to one of lower pressure.

Juvenile Hormones (for Insect Control): Natural or synthetic compounds which have growth regulatory effect on the juvenile stages of insects, *i.e., on* larvae and pupae, and not the adult. Many of them are known to inhibit the metamorphosis of the larva to pupal or adult stage and are used for insect control, *e.g.* dehydiojuvabione against *Pyrrhocoirs* apterous and hydroprene, methoprene, triprene and kinoprene against hemlock looper *(Lambadina fiscellaira),* A natural juvenile hormone (JH) is produced by an insect in the process of its immature development which maintains its nymphal (pupal) or larval form.

K

Kainite: A double salt of magnesium sulphate and potassium chloride $MgSO_4$. $KCl.3H_2O$, that occurs naturally in Poland and in the Stassfurt Deposits. A valuable source of potassium salts.

Kaolin: Fine white clay called China clay and porcelain clay. Widely used for making pottery, as a coating for paper and as a filler for rubber tyres.

Kari Fischer Reagent: A mixture of I_2 and SO_2 dissolved in pyridine MeOH used as a titrant for water with which HI is liberated and the pH determined with a meter.

Kekule Structure of Benzene (C_6H_6): The structural formula for benzene as a simple hexagonal ring of six interconnected carbon atoms. Alternate single and double bonds are present in the same ring.

Kelvin: A unit of temperature also known as absolute unit. In the kelvin scale absolute zero is 0 K and is equal to –273.16°C, the freezing point of water is 273 K and boiling point is 373 K.

Kepone ($C_{10}Cl_{10}O$): A highmelting ketonic solid obtained by condensation of two molecules of hexa-chloro-cyclo-pentadiene in the presence of SO_3, oleum or HSO_3Cl. It is used as insecticide and fungicide.

Kernite or Resorite: A hydrated sodiumborate, $Na_2B_4O_7.4H_2O$ which is found in California and is a major source of boron compounds.

Ketal: A kind of organic compound obtained by addition of an alcohol to a ketone.

Ketene (CH_2:C:CO): A colourless gas, b.p. –56°C, used as an acetylating agent in the manufacture of cellulose acetate and aspirin. It is the first member of the ketene series, which has the general formula

R:C:CO, where R represents a bivalent radical or two univalent radicals.

Ketoenol Tautomerism: The type of tautomerism that occurs in ketones as the result of the migration of a hydrogen atom from an alkyl group of the carbonyl group. Thus, acetone contains in addition to ketone molecules ($CH_3.CO.CH_3$, the ketoform). A small proportion of molecules having the structure of an unsaturated alcohol ($CH_2:COH.CH_3$, the enolform).

Ketohexose: A ketose sugar with six carbon atoms.

Ketone: A carboxyl group is bonded to two alkyl groups which may be alike or different.

Kieselguhr: Diatornaceous earth, infusorial earth. A mass of hydrated silica (SiO_2) formed from skeletons of minute plants known as diatoms. It is a very porous and absorbent material, used for filtering and absorbing various liquids, in the manufacture of dynamite and in other industries.

Killed Spirits of Salts: A solution of zinc chloride, $ZnCl_2$, made by reacting zinc with hydrochloric acid. Used in soldering.

Kiln: A furnace for making bricks, ceramics and beating 'chalk to form lime and carbon dioxide.

Kiloton Bomb: A nuclear weapon with an explosive power equivalent to one thousand tons of T.N.T.

Kilowatt-hour (Symbol kwh): A unit of energy, usually electric energy equal to the energy transferred by one kilowatt of power in one hour. It has a value of 3.6×10^6 joules.

Kindling Temperature: The lowest temperature at which a material catches fire, and burns.

Kinetics: A branch of physical chemistry that deals with the study of rates of chemical reactions and the effect of physical conditions that influence the rate of reaction. The measurement of these rates will out the mechanism of the reaction.

Kinetic Theory of Gases: A theory which explains the behaviour of gases under various conditions. The theory says that gas particles in continuous motion bombard the walls of the container producing pressure. When temperature is raised speed of particles increases and so does the pressure, If volume in reduced there are more particles bombarding unit area of walls and so does pressure increases.

Kish: A variety of graphite occasionally formed in iron smelting furnaces.

Kjeldahl Flask: A round-bottomed glass flask with a long wide neck, used in the estimation of nitrogen by Kjeldahl's method.

Kjeldahl's Method: A quantitative method to estimate nitrogen from the organic compounds, It is converted into ammonium sulphate when heated strongly with concentrated sulphuric acid.

Knocking in the Internal-Combustion (petrol) Engine: Violent explosions in the cylinder, often due to over-compression of the mixture of air and petrol vapour before sparking.

Kohlrausch Equation: An equation which tells the behaviour of strong electrolytes on dilution.

$$\Lambda_\infty - \Lambda_v = kc^{1/2}$$

where Λ_∞ is the equivalent conductivity at infinite dilution, that a volume v, k is a constant and c is the concentration of electrolyte. This equation is valid only at high dilution.

Kolbe's Electrolytic Reaction: A process for preparing alkanes by the electrolysis of sodium salt of carboxylic acid.

Kolbe's Reaction: An industrial method to prepare salicylic acid by heating sodium phenoxide at 180-200°C in a current of CO_2 and the product obtained is treated with acid.

Kovar: An alloy of cobalt, iron and nickel, which has a coefficient of expansion similar to that of glass. Used for glass-to-metal seals, particularly in thermionic valves and transistors.

Kroll Process: The process of reduction of metal halides by magnesium, sodium or calcium, e.g., tinatum can be formed by reduction of $TiCl_4$ with magnesium

$$TiCl_4 + 2Mg \rightarrow 2MgCl_2 + Ti$$

Kryptol: A mixture of graphite, carborundum, and clay, used as an electrical resistance in electric furnaces.

Kurchatovium: Artificially produced radioactive element belonging to group 4. Also known as rutherfordium but now being called unnilquadium.

L

Lachrymators: Volatile substances that cause an irritating effect on the eyes and produce tears, e.g. mustard gas or 2, 2-dichlorodiethyl sulphide, $CICH_2CH_2SCH_2CH_2CI$.

Lactalbumin: A protein belonging to the albumin class present in milk. It has a molecular weight of about 17,500.

Lactam: A type of organic compound containing the —NH.CO group as a part of arising in the molecules, They can be regarded as formed from a straight chain compound that has an amine group at one end of the molecule and a carboxylic acid group of the amine group with the carboxylic acid group, with elimination of water, leads to cyclic attractive which is thus an internal amine.

Lactams: Obtained when aminoacids heated to lose water. They are decomposed by heating with alkalis. They are solid and are very poisonous.

Lactate: A salt or ester of bacteria acid.

Lactides: Obtained when a-hydroxy-fatty are heated they lose water. They are decomposed by heating with water to give the original acids.

Lactone: A type of organic compound containing the soda group —O.CO as part of a ring in the molecule. A lactone can be regarded as formed from a compound with an alcohol (OH) group on one end of the chain and a carboxylic acid (COOH) group on the other, then results from reaction of the OH group, i.e. It is an internal ester.

Lactose ($C_{12}H_{22}O_{11}$): A sugar found in milk. It is a disaccharide composed of glucose and sodalactose units.

Lambda Point: The temperature of transition between the He_i and He_{11}

Lambert-Beer's Law: The relationship e = A/cl (a combination of Beer's and Lambert's Laws that forms the basis of absorption spectrometry) where e is the molar extinction coefficient (molar absorption coefficient or molar absoiptivity), *c* the molar concentration of solute, 1 the pad, length and A the absorbance. Results are often reported in terms of molar absorptivity e or its logarithm. When molecular weight of a substance (solute) is unknown, of absorption is conveniently expressed as $A^{1\%}_{1\ cm}$ value, i.e. the of a 1% solution of the 1 cm substance in a 1 cm cell. Ibis value is easily related to the molar absorptivity by the expression 10 e = $A^{1\%}_{1\ cm}$ ´ molecular weight. 1 cm

Lamberta Law: States that layers of equal thickness of a homogeneous material absorb proportions of light.

Laminated iron: Thin sheets or iron used for cores of transformers instead of solid iron core, in order to reduce losses due to eddy currents,

Lamp Black: A soft black carbon pigment prepared by incomplete combustion of natural gas or petroleum. It is used in paints and inks.

Lanthanum: A soft ductile malleable silvery metallic element that is first member of the lanthanoid series. It is found associated with other lanthanides in many minerals, including monazite, and bartnase Lanthanum is used in several alloys as a catalyst and in compound form in carbon are searchlights and in the glass industry.

Lanthanides: A group of element (Ce, Pr, Nd, Pm, Sm, Eu, Gd, Tb, Dy HO, Er, Tm, Yb, Lu) whose electronic configurations display back filling of the 4f level (except for lanthanum) and hence have been grouped (in group III) along with lanthanum in the conventional periodic table of elements. All these elements have oxidation state' M^{3+} and great similarity in the size of the ions and chemical properties.

Lanthanide Contraction: The lanthanides (elements 58 to 711 show a Slow but a significant decrease in atomic rates. The, differentiating electrons of these elements are added (0 the third level from the outside (the 4f subshell). These electrons screen the increasing nuclear charge from the outer 6S electrons but are not completely efficient and hence the steady reduction in size is observed.

Lanthanide Series: The elements with atomic numbers 57,71 in which the 4f subshell is being built up, constitute the lanthanide (rare earth element) series, These elements are very much alike in chemical and physical properties. The principal oxidation of these elements is + 3.

Lanthanides: A group of elements whose electronic configurations display back 01 the 4R-level, There is a maximum of 14 electrons in an F orbital. The element lanthanum, itself has no F-electrons and is thus strictly not a lanthanoid, but it is included by convention, thus giving related series a closely related series of 15 elements.

Lassaigne's Test: A common test for the detection of nitrogen, halogens or sulphur in an organic compound, a little of which is heated with a pellet of sodium in a hard glass test tube, the hot tube is plunged into distilled water and the product ground. The presence of nitrogen is shown by the formation of Prussian Blue on heating with a solution Iron (II) Sulphate containing a trace of Fe (III) and hydrochloric acid. With halogen containing substance, sodium halide is obtained, delectable with silver nitrate, while the presence of sulphur can be shown with sodium a nitroprusside.

Latent Heat: The heat required by a substance or a system without an accompanying rise in temperature during a change of state from solid to liquid or liquid to gas.

Lattice: (1) The regular network of fixed points about which molecules, atoms, or ions vibrate in a crystal. (2) In a nuclear reactor, a structure consisting of discrete bodies of fissile and non-fissile material (especially moderator), arranged in a regular geometrical pattern.

Lattice Energy: The energy released during the formation of one mole of a crystal of the solid from the constituent gaseous ions separated from each other by infinite distances.

Law of Mass Action: The rate at which a substance reacts is proportional to its active mass and the rate of a chemical reaction is proportional to the product of the active masses of the reacting substances.

Law of Multiple Proportions: When two element A and B combine to form more than one compound, the weights of B that combine with a fixed weight of A are in small whole number ratio, e.g. nitrogen and oxygen form oxides namely nitrous oxide, N_2O; nitric oxide, NO; nitrogen peroxide N_2O_4 and nitrogen pentoxide, N_2O_5; the amounts of nitrogen that combine with a fixed weight of oxygen are in the ratio of 5 : 4 : 2 : 1.

Lawrencium: A radioactive transuranic element of actinide series obtained by bombardment of Californium with accelerated ^{10}B ions. Symbol : Lr; a.n. 103.

LCAO Method: A way of calculation of molecular orbital based upon the concept that the molecular orbital can be expressed as a linear combination of the atomic orbital.

LD50: Median lethal dose — the dose of toxin at which 50% of exposed animals are killed, It is used as a standard measure of toxicity of drugs.

Lead: A silvery white metal belonging to group 14, It is obtained by the reduction of roasted ore with coke. The main source of lead is galena, Pbs. It quickly tarnishes due to the formation of oxide and carbonate.

Lead Accumulator: A cell which can be recharged after it is discharged. It consists of two plates of lead, one filled with finely divided lead (negative plate) and ether lead dioxide (positive plate). Both of these arc dipped in a solution of 10-15 percent of sulphuric acid.

Lead Arsenate [$Pb_3(AsO_4)_2$]: A white crystalline substance, m.p. 1042°C, used as an insecticide.

Lead Bromide ($PbBr_2$): A white precipitate obtained by the addition of a soluble bromide to the solution of lead salt.

Lead (H) Carbonate ($PbCO_3$): A white solid exists in nature as cerussite. It is obtained as white crystalline precipitate by passing carbon dioxide into a solution of lead acetate

$$(CH_3COO)_2Pb + H_2O + CO_2 \rightarrow PbCO_3 \downarrow + 2CH_3COOH$$

Lead (II) Carbonate Hydroxide ('White Lead): A white powder obtained when perforated discs react with 3% acetic acid in earthern pots stacked in a row over horse dung. The pots arc allowed to stand for 68 weeks during which interval acetic acid vapour, moisture and carbon dioxide ail act upon lead and give flaky white lead. It is used as white paint.

Lead-Chamber Process: The manufacture of sulphuric acid by the action of nitrogen dioxide, NO_2, on sulphur dioxide, SO_2, to give nitric oxide, NO, and sulphur trioxide, SO_3. The former reacts with oxygen from the air to give NO_2 again; the SO combines with water to give sulphuric acid, the process being carried out in large lead chambers.

Lead Chloride ($PbCl_2$): A white precipitate obtained when hydrochloric acid or some soluble chloride is added to a lead solution

$$Pb(NO_3)_2 + 2HCI \rightarrow PBCl_2 \downarrow 2HNO_3$$

Lead Chromate ($PbCrO_4$): A yellow precipitate obtained after mixing solution of lead salts and potassium chromate. It is used as a yellow pigment.

Lead (II) Sulphate ($PbSO_4$): A white solid, exists in nature as anglesite. It is obtained by adding dilute sulphuric acid or a soluble sulphate to a solution of lead salt. It is insoluble in water but partially dissolves in hot concentrated sulphuric acid and separates out or on cooling. It is used in gravimetric estimation of lead.

Lead (II) Sulphide ($PbSO_4$): A black Precipitate obtained by passing hydrogen sulphide gas through a solution of a plumbous salt. It occurs in nature a galena. It dissolves in nitric acid. It is insoluble in yellow ammonium sulphide.

Lead Tetraethyl [TEL(C_2H_5)$_4$Pb]: A colourless liquid obtained by heating ethylchloride with sodiurnlead alloy under pressure. It is insoluble in water but soluble in ether. It is very poisonous and needs special precautions in handling. It is used as an antiknock compound which improves to quality of Petrol.

Lean Gas: A fuel gas defined as containing a calorific value of 100-450 Btu/ft^3 (3.716.8 mJ/m^3), obtained from liquid and solid hydrocarbons by steam refining in the presence of a catalyst at about 1075 K to give a mixture of CO and H_2O.

Leblanc Process: An obsolete process for the manufacture of sodium carbonate, sodium chloride is converted to sodium sulphate by heating with sulphuric acid. This is then roasted in a rotary furnace where it is first reduced to the sulphide using carbon and then immediately converted to the carbonate solution is obtained by leaching with water and this is subsequently dried and calcium to obtain the solid.

Le Chatlier's Principle : If one of the factors of a system in equilibrium (pressure temperature or concentration) is altered, one or more of the other factors will spontaneously adjust themselves in such a way as to reduce the effective alteration.

Lepton: It is a fermian size types of leptons are electrons, muoans, taus, and three kinds of neutrines.

Levis Acid: This is a substance that can accept an electron pair to form a coordinate bond; a levels base is an electron pair donor.

Libermann's Reaction: A colour test for —NO or for the detection of an —NO group. A small quantity of the substance is dissolved in concentrated sulphuric acid and a crystal of phenol added. A blue colour forms on warming when the test is used for detecting a phenol, the substance and a crystal of sodium nitrite arc dissolved in sulphuric acid and warmed. On dilution and addition of alkali many phenols give distinctive colours.

Liebig Condenser: A simple type of laboratory condenser. It consists of a straight glass tube, in which the vapour condensed with a surrounding glass jacket through which cooling water fibrous.

Ligand: A molecule or ion that is capable of giving a pair of electrons and forms a coordinated bond.

Light (Visible Radiation): A form of electromagnetic radiation above to human eye. Its wavelength range is between 400 nm and 700 nm.

LimeSoda Process: A method for removal of Mg^{2+} and Ca^{2+} ions from water to reduce hardness of water with the addition of lime, $Ca(OH)_2$ and soda ash, Na_2CO_3.

Lime Water: A solution of calcium hydroxide, $Ca(OH)_2$, in water. It turns milky by the action of carbon dioxide, CO_2, owing to the formation of insoluble calcium carbonate, $CaCO_3$.

Line Spectrum: A spectrum containing radiation of only specific wavelength, e.g., hydrogen spectrum.

Lipowitz Alloy: A fusible alloy, m.p. 65. 70°C, consisting of 50%, bismuth; 27% lead, 13% tin and 10% cadmium.

Liquation Process: A process employed when the impurity is less fusible than the metal itself. The impure metal is placed on the sloping hearth of a furnace and gently heated. The metal melts and drains away leaving behind the infusible materials on the hearth.

Liquid: The physical state of matter in which the particles of a substance are loosely bound by intermolecular forces.

Liquid Air: A pale blue liquid, It is a mixture of liquid oxygen (boiling at –183°C) and liquid nitrogen (boiling at –196°C).

Liquid Crystats: A substance which on heating gives cloudy liquid, e.g., p-azoxyanisole and ammonium oleate.

Liquid Natural Gas (LNG): Natural gas liquefied at atmosphere pressure by cooling to about –160°C. It is for the purpose of transportation.

Liquid Petroleum Gas (LPG): A gas liquefied by compression, consisting of flammable hydrocarbons like propane and butane, obtained as a by product from refining of petroleum or from natural gas. It is used as domestic and industrial fuel.

Lithium: A silver white metal belonging to group 1. It is electrolysis of a mixture of sodium chloride and potassium chloride. It is highly reactive though less so than the rest of the alkali metals. It is used in the manufacture of various alloys and used in producing thermonuclear energy.

Lithium Hydroxide (LiOH): Anhydrous lithium oxide obtained on heating the hydrated crystal of LiOH H_2O in a current of hydrogen. LiOH H_2O is obtained by the action of baryta water on lithium sulphate

$$Li_2SO_4 + Ba(OH)_2 \rightarrow BaSO_4 \downarrow + 2LiOH$$

Lithium Hydrogen Carbonate (LiHCO3): A compound known only in solution, produced by passing carbon dioxide through a suspension of lithium carbonate

$$Li_2CO_3 + H_2O + CO_2 \rightarrow LiHCO_3$$

The solution is known as lithia water and used in medicines for treatment of goat.

Lithium Nitrate ($LiHO_3$): Obtained by dissolving the carbonate in nitric acid

$$LiCO_3 + 2HNO_3 \rightarrow 2LiNO_3 + H_2O + CO_2$$

Lithium Nitrate ($LiHCO_3$): Lithium combines with nitrogen even at ordinary temperature and more readily on heating in a current of nitroger

$$6Li + N_2 \rightarrow 2Li_3N$$

It reacts with water to give lithium hydroxide and ammonia.

$$Li_3N + 3H_2O \rightarrow 3LiOH + NH_3$$

Lithium Oxide (Li2O): A white powder prepared by healing the lithium in air.

$$4Li + O_2 \rightarrow 2LiO_2$$

Lithium Oxide (Li_2O): A white crystalline precipitate obtained on adding H_2O_2 and alcohol to a solution of lithium hydrogen. When dehydrated over P_2O_5, it yields the anhydrous lithium peroxide.

Lithium Phosphate (Li_3PO_4): A precipitate obtained on adding a soluble phosphate to the solution of a lithium salt or by The action of phosphoric minerals acids and the solution on evaporation gives soluble acid phosphate, LiH_2PO_4.

Lithium Sulphate (Li_2SO_4): A white solid excess of lithium oxide or carbonate to sulphuric acid. It is very much soluble in water.

Lithium Tetra Hydriodo Aluminate (M) (Lithium Aluminium Hydride, $LiAlH_4$): A white solid obtained by the action of anhydrous aluminium chloride on lithium hydride in ethereal solution.

$$4LiH + AlCl_3 \rightarrow LiAlH_4 + 3LiCl$$

It is powerful reducing agent.

Lithophone ($BaSO_4 + ZnSO_4$): A soft white amorphous product used as pigment. It has very good covering powder and is not discoloured by H_2S and other sulphurous gases present in the atmosphere of big cities.

Litmus: A soluble purple substance prepared from plants. In acid medium it turns red; in alkaline medium, it turns blue. It is used to find out the end point in the of acid and base.

Litre (Symbol l): A unit of volume now defined as 1000 cm^3

Lixiviation: The process of separating soluble compounds from a mixture by washing them out with water.

Localized Bond: A bond in which the electrons contributing to the bond remain between the two atoms concerned, i.e., the bonding orbital is localized. The vast majority of bonds are of this type.

Lone Pair: A pair of electrons in a molecule which is not shared by two of the constituent atoms, i.e., does not involve in direct bonding. It forms coordinate bond, e.g., NH_3 (one lone pair) and OH_2 (two lone pairs).

Loschmidt's Number: The number of molecules of an ideal gas in unit at STP; equal to 2.687×10^{19} per cm^3.

LSD: Lysergic Acid Diethylamide. An extremely powerful drug that causes distortions in thinking and feeling.

Lunar Caustic (Silver Nitrate, $AgNO_3$): Usually fused and cast into sticks.

Lustre (Metallic Luster): The electrons readily absorb light falling upon them, where upon they move to a higher level but fall back to lower one immediately. This is accompanied by the emission of the light absorbed. Thus, most of the light is readily reemitted, causing the metallic lustre.

Lutetune: A silvery element from the lanthanide series of metals. It occurs in association with other lanthanoids.

Lyddite: An explosive consisting of picric acid, mixed with 10% nitrobenzene and 3% vaseline.

Lyman Series: A second series of lines in the ultraviolet region. The frequencies may similarly be expressed as *1/n* times a series of difference between energy terms in which *n* has the value of 1 in the constant term and takes the successive integral value 2, 3... in the variable term.

Lysergic Acid ($C_{15}N_{15}N_2COOH$): A crystalline substance obtained from ergot and used in the manufacture of the hallu nitrogen LSD.

M

Macrolides: Compounds with antibiotic activity obtained by streptomycees spp. and characterized by having a macrocyclic ring.

Macromolecular: A large molecular; e.g., a natural or synthetic polymer.

Macromolecule: A molecule with hundreds Lo thousands of atoms such as found in polymers.

Macromolecular Crystal: A crystal composed of atoms joined together by covalent bonds, which form giant three dimensional or two dimensional networks. Diamond is an example of a macromolecular structure.

Magic Numbers: The numbers 2, 8, 20, 28, 50, 82 and 126. Atomic nuclei containing these numbers of neutrons or protons have exceptional stability.

Magma: A hot, sticky liquid state of rock made by conditions of high temperature and high pressure.

Magnalium: A light alloy, r.d. 22.5. It consists of aluminium with from 5% to 30% magnesium.

Magnesium: A silvery white metal which soon becomes dull in air. This belongs to IIA group (alkaline-earth metals). It is moderately hard, fairly ductile and rolled into thin sheets. It exists in nature as magnesite, $MgCO_3$; dolomite, $MgCO_3.CaCO_3$; carnallite, KClMg $Cl_2.2H_2O$ Epsom salt, $MgSO_4.7H_2O$, talc or soap stone, $Mg_3H_2(SiO_3)_4$. It is extracted by the electrolysis of fused magnesium chlorides. It is used in flash bulb with potassium chlorate photography, in the formation of its alloys, pyrotechny and for igniting thermite in alumino thermic processes, extraction of silicon by ignition of silica and the formation of reagent 15 largely used for the synthesis of many organic compounds.

Symbol : Mg; a.n. 12; m.p. 641°Cb.p. 1107°C; r.d. 1.74.

Magnesium Bromide (MgBr2): A white crystalline substance obtained by heating magnesium in bromine vapour. It is also found in sea water.

Magnesium Carbonate ($MgCO_3$): A white powder, which occurs in nature as magnesite, $MgCO_3$ and dolomite, $MgCO_3.CaCO_3$. It is used in medicine as a mild antacid.

Magnesium Chloride ($MgCl_2$): The anhydrous salt can be obtained by heating the hexahydrate in a current of dry HCI gas. Whereas hydrated magnesium chloride is obtained by dissolving magnesium carbonate or oxide in dilute hydrochloric acid followed by crystallization. The crystalline magnesium chloride is a hexahydrate and is very deliquescent. It is used as lubricant for thread, in spinning and weaving industry. A paste of a strong solution of magnesium chloride with magnesium oxide sets into a stone like mass owing to the formation of the oxy chloride. This mixture commercially known as Sorel's cement is used for making artificial stone and as a cement for teeth.

Magnesium Hydroxide [($Mg(OH)_2$)]: A white powder, only slightly soluble in water, It is obtained as a white gelatinous precipitate by adding in to the solution of any soluble salt of magnesium, e.g.,

$$MgSO_4 + 2NaOH \rightarrow Mg(OH)_2 \downarrow + Na_2SO_4$$

It has an alkaline reaction and is used in the sugar industry for the extraction of sugar from molasses.

Magnesium Iodide (MgI_2): A white hygroscopic crystal obtained by dissolving magnesium oxide in hydxoiodic acid. It is an unstable substance and easily decomposes liberating iodine,

Magnesium Nitride (Mg_3N_2): A greenish-yellow amorphous powder obtained by burning magnesium in the atmosphere of nitrogen. It reacts with water forming magnesium hydroxide and ammonia.

$$Mg_3N_2 + 6H_2O \rightarrow 3Mg(OH_4)_2 + 2NH_3$$

Magnesium Oxide (Magnesia, MgO): A white powder, soluble in water. A paste of magnesia and water set like a mortar, due to dehydration and absorption of carbon dioxide from the atmosphere. It is a basic oxide and dissolves in acids to form magnesium salts. It is obtained by burning magnesium in air

$$2Mg + O_2 \rightarrow 2MgO$$

It is used for the refractory lining of furnaces and for making crucibles and cupels used for metallurgical processes. It is also used in medicine for correcting acidity.

Magnesium Perchlorate [Mg(ClO$_4$)2]: Commercially obtained by dissolving magnesium hydroxide in perchloric acid, $HClO_4$

$$Mg(OH)_2 + 2HClO_4 \rightarrow Mg\ (ClO_4)_2 + 2H_2O$$

It is used as a drying agent.

Magnesium Peroxide (Hg_2O_2): A white insoluble solid prepared by reacting sodium or barium peroxide with a concentrated solution of a magnesium salt. Used as a bleach for dyestuffs and silks.

Magnesium Sulphate ($MgSO_4$): In nature exists as kieserite ($MgSO_4.H_2O$) in Stassfurt deposits or as Epsom salt in the mineral water of the Epsom springs in England. The salt $MgSO_4.H_2O$ exists in well defined transparent crystals which lost water of efforescence. It is fairly soluble in water and gives acidic solution due to hydrolysis. It is used as purgative in medicine, as mordant in dyeing, in the manufactures of paints, soaps and as catalyst carrier in the manufacture of sulphuric acid in Grillos' converter.

Magnesium Sulphide (MgS): Obtained by the action of hydrogen sulphide on magnesium nitride.

$$Mg3N_2 + 5H_2S \rightarrow 3\ MgS + 2NH_4HS$$

Magnesium Trisilicate (Dimagnesium Trisilicate, 2MgO. 3SiO$_2$.nH$_2$O): A white tasteless powder used as an antacid and to absorb odours.

Magnetic Moment: The magnetics moment (m) of the magnetic substance, is related to the magnetic susceptibility (c) by the equation

$$\mu = 2.84\ \sqrt{\chi_m}\ (T - \theta)$$

where χ, is the molar susceptibility, T the absolute temperature and is a constant.

Magnetic Separation: A method of separating Crushed mineral mixtures using the magnetic properties which a component may possess, such as ferromagnetism or diamagnetism.

Magnetic Tape: Plastic tape coaled a powder, used in tape words. The tape is passed over the gap in a magnetic circuit, which is modulated in accordance with to be recorded. The tape retains a record of the modulation, which can be 'played back' through a suitable circuit. Magnetic tape is used in the backing storage of computers.

Magneton: A unit for measuring the magnetic moments of atomic particles.

The Bohr magneton, *u*E, is equal to

$$eh/4rm_e = 9.2741 \times 10^{-24}$$

where e and m_e are the charge and mass of the electron and h is Planck's The nuclear Magneton μN, is equal to

$$\mu p.m_e/m_p = 5.05 \times 10^{-27} \text{ ampere metre}^2$$

where m_p is the mass of the proton. The symbols mB and m_N are sometimes used for the Bohr magneton and the nuclear magneton respectively.

Magnox: A magnesium alloy used for sheathing uranium fuel elements in certain types of nuclear reactor.

Malachite: A green mineral consisting of copper (II) carbonate and hydroxide [$CUCO_3.Cu(OH)_2$]. it is used as an ore and a pigment.

Malt: Grain (usually barley) which has been allowed to germinate and then heated and dried.

Maleic Acid ($C_4H_4O_4$): A colourless crystalline solid (m.p. 403 K) soluble in water. It is used in the formulation of polymers.

Maleic hydrazide or 3, 6 dioxo-1, 2, 3, 6 tetrahydropyridazine C4H4N2O2: A colourless crystals obtained by condensation of anhydride with aqueous hydrazine It is used to retard the growth of some plants.

Malic Acid or Monohydroxysuccine Acid (HOOC. CH_2.CHOH.COOH): A colourless crystalline and solid, readily soluble water and alcohol it is gradually replacing citric acid used in is also used as a remedy for sorethroat and as, a purgative.

Malleability and Ductility: A metallic bond is cohesive but not when a metal is hammered into different shapes, it is merely necessary for the atom to assume different positions.

Manganese: A transition metal occurring naturally as oxides, e.g., pyrolusite (MnO_2). It forms very useful alloy with iron. It is also used as a dioxidiser, desulpnurise and dephosphoriser for steel and other metals.

Symbol : Mn; a.n. 25; m.p. 1244°C; b.p, 2040°C; r.d. 7.4.

Manganese (II) Chloride (Manganous Chloride $MnCl_2$. H_2O): A pink monoclinic crystal of hydrated salt ($MnCl_2.4H_2O$) obtained by dissolving the metal oxide hydroxide, or carbonate in hydrochloric acid. The anhydrous salt is prepare by heating manganese in chlorine. It is very deliquescent and highly soluble ill water. It is used as catalyst and in the preparation of other compounds of manganese.

Manganese (II) Hydroxide or Manganous Hydroxide [Mn (NOH)$_2$]: A white precipitate formed when an alkali solution is added to a solution of manganous salt

$$MnCl_2 + 2NaOH \rightarrow Mn(OH)_2 + 2NaCl$$

It rapidly changes brown in air owing to its oxidation.

Manganese (II) Nitrate or Manganous Nitrate [Mn(NO$_3$)$_2$]: Obtained by dissolving manganous carbonate in dilute nitric acid. Hexahydrate crystals of Mn (NO_3) are colourless deliquescent crystals.

Manganese (II) Oxide (Manganic Oxide MnO): A greyish green powder obtained by reducing a higher oxide with hydrogen or carbon monoxide at 120°C. It is insoluble in water.

Manganese (III) Oxide (Manganic Oxide, Manganese Sesquioxide, Mn_2O_3): A black powder obtained by heating manganese dioxide in a stream of ammonia. It reacts with cold dilute acid to form manganic salts.

Manganese (IV) Oxide (Manganese Dioxide, MnO_2): A black powder insoluble in water. It decomposes on heating into Mn_3O_4 and O_2. It is used as catalyst, in the preparation of chlorine gas, in preparing dry cells and used in glass industry for bleaching yellowish colour due to iron oxide.

Manganese Steel: Very hard variety of steel containing up to 13% manganese.

Manganese (II) Sulphate or Manganous Sulphate ($MnSO_4$): Obtained by heating commercial black dioxide of manganese with sulphuric acid to redness.

$$2MnO_2 + 2H_2SO_4 \rightarrow 2MnSO_4 + 2H_2O + O_2$$

Manganin: An alloy containing 83% copper, 13% manganese, 4% nickel. As its electrical resistance is affected only slightly by change in temperature it is used in resistance Coils.

Mannitol Hexanitrate [$C_6H_{12}(ONO_2)_6$]: A clourless insoluble substance, m.p. 112°C, used as an explosive and in medicine.

Mannose ($C_6H_{12}O_6$): A simple sugar found in many polysaccharides. It is an aldohexose, isomeric with glucose.

Manometer: A device for measuring pressure. In its simplest form it contains a U-tube (each arm at least 76 cm long half filled with mercury. When pressure/vacuum is applied to one end of the tube, the difference in the heights of mercury column in the two arms gives the Pressure difference.

Markownikoff's Rule: The negative part of the reagent adds on to the carbon atom that is joined to the least number of hydrogen atom.

Marsh Test for Aresenic: The Aresenic containing compound is changed to volatile AsH_3 which is decomposed to brown stain on heating. Estimation by comparison of stains.

Martensite: The hard and brittle constituent of steel produced when the material is cooled from its hardening temperature at a greater rate than its critical cooling rate.

Maser: An acromyne for Microwave Amphification by stimulated Emission of Radiation.

Mass (Symbol m): A measure of the quantity of matter in an object. The SI unit of mass in the kilogram.

Mass Defect: The difference between the mass of a nucleus arid the sum of the masses of its constituent nucleons, The energy equivalent of the mass defect, on the basis of the mass-energy equation, in must be supplied to a nucleus to split it into its component nucleons.

Mass-Energy Equation: The equation $E = mc^2$ where E is the total energy of a mass m, c being the velocity of light.

Mass Spectrograph Aston's: An instrument used to identify isotopes. By the successive application of electric arid magnetic field he got series of impressions on the photographic plate, each of which represent a different isotope.

Matches: The heads of safety matches usually contain antimony trisulphide, oxidizing agents, such as, potassium chlorate, and some sulphur or charcoal; while the striking surface contains red.

Matte: A mixture of copper and iron sulphides obtained at an stage in smelting copper ore.

Matter: The physical material of the universe which occupies space has mass.

Matter Wave : The wave characteristics or subatomic

Maxwell: Symbol Ms. A unit of' magnetic flux used in the e.g,s, It is equal to 10^{-8} wb.

Maxwell-Boltzmann Distribution: A statistical equation giving the distribution of velocities or positions of the molecules in a gas; it is based on the assumption that all the particles are indistinguishable and each has an equal probability of appearing in a particular region.

McCleod's Equations

$$g = K(D-d)^4$$

where is the surface tension, D is the density of the liquid and d the density of the vapour.

Mean Free Path: The average distance covered by a molecule between collisions with another molecule. At STP the mean free path for 1.8 10^{-6}

Mechanism: A step by step description of the events taking place in a chemical reaction. It is a theoretical framework accounting the fate of bonding electrons and illustrates which bonds are broken and which are formed. For e.g., in the chlorination of methane to give chloromethane.

Step 1: Cl:Cl $\rightarrow$ Cl

Step 2: Cl + $CH_4 \rightarrow$ HCl + CH_3

Step 3: CH_3 + Cl:Cl $\rightarrow$ CH_3C1 + Cl

Median Lethal Dose: The dose of ionizing radiation which would kill 50% of a large batch of organisms a specified period.

Median Survival Time or ST_{50}: A restricted concept of median effective time, the time at which death occurs in half the test subjects, after exposure to a pathogenic (including toxicological) stimulus. ST_{50} is not a direct measure of dosage, unlike LT_{50}.

Meerchaum: Natural hydrated magnesium silicate. White solid used for tobacco pipes.

Megahertz MHz: (1 million hertz). A measure of frequency equal to 10^6 cycles per second.

Megaton Bomb: A nuclear weapon with an explosive power equivalent to one million tons of T.N.T.

Melamine (C3H6N6), Triminotriazine: A white crystalline substance, m.p 354°C, that forms a thermosetting resin with formaldehyde.

Melanin: A polymeric pigment responsible for colouring of eyes, skin and hair. Formula is $C_{17}H_{98}O_{33}\ N_{14}S$.

Mendeleev's Periodic Law: A Russian chemist in 1869, suggested that the properties of the elements are periodic function of their atomic mass.

Mendius Reaction: The alkyl cyanide reduced to alkyl amine in the presence of lithium aluminium hydride ($LiAlH_4$) or sodium in alcohol.

Meniscus: A concave or convex upper surface that forms on a liquid a tube as a result of surface tension. Water forms a concave meniscus while mercury forms a convex one.

Menthol ($C_{10}H_{19}OH$): A colourless, volatile, prismatic crystal (m.p. 318 K), having strong peppermint odour and a cooling taste. It is used as antiseptic in ointments, in medicine and in perfume industry.

Mercuric Fulminate: Mercuric isocyanates $H_5(ONC)_2$ A white crystalline substance which explodes on being struck, used in detonators to initiate explosions.

Mercury: A silver white liquid metal obtained by roasting and distillation of Cinnabar (HgS). It is not attacked by oxygen or air at ordinary temperature. It is used for filling thermometers, manometres, vacuum pump, diffusion pump etc. Mercury lamps are used as source of ultraviolet light. Mercury compounds are used in medicines.

Symbol . Hg; a.n. 80; m.p. 61°C; b.p. 357°C; r.d. 13.5.

Mercury (1) Carbonate or Mercurous Carbonate (Hg_2CO_3): A yellow precipitate obtained when potassium bicarbonate solution is added to mercurous nitrate solution. It decomposes on heating (400 K) or on exposure to light.

$$Hg_2CO_3 \rightarrow HgO + Hg + CO_2$$

Mercury (II) Carbonate or Mercuric Carbonate ($HgCO_3$): Obtained by treating a large excess of potassium carbonate and potassium bicarbonate solution respectively with mercuric nitrate solution. It is known only in the form of basic carbonate $HgCO_3.2HgO$ (yellow) and $HgCO_3.3HgO$ (brown).

Mercurodrome: A red dye obtained by mercuration of dibromofluorescein followed by treatment with Caustic Soda. Its 2% solution is used as an antiseptic.

Mercury Cell: A primary cell consisting of a zinc anode, a cathode of mercuric oxide (HgO) mixed with graphite (about 50%), and an electrolyte of potassium hydroxide (KOH) saturated with zinc oxide (ZnO). The E.M.F. is about 1.3 volts and by suitable design the cell can be made to deliver about 0.3 ampere-hour per cm^3.

Mercury (I) Chloride (Calomel or Mercurous Chloride, Hg2Cl2: A white powder without any taste or smell. It is insoluble in water. It is prepared by mixing a solution of a chloride with a soluble mercury (I) salt.

$$Hg_2^{2} + 2Cl \rightarrow Hg_2Cl_2$$

It gets blackened on treating it with ammonia due to the formation of a compound of the formula $H_2N–Hg–Hg–Cl$

$$Hg_2Cl_2 + 2NH_3 \rightarrow 2HN–Hg–Hg–Cl + NH_4C1.$$

This compound is unstable and changes into mercury and mercuric amino chloride

$$H_2N–Hg–Hg–Cl \rightarrow H_2N–Hg–Cl + Hg$$

It is used in medicine as a purgative and preparing calomel electrode.

Mercury (II) Chloride (Mercuric Chloride, $HgCl_2$): A white crystalline solid which sublimes on heating. It is prepared when a mixture of mercuric sulphate and common salt is heated with a little of manganese dioxide.

$$HgSO_4 + 2NaCl \rightarrow HgCl_2 + Na_2SO_4$$

When treated with ammonia, it forms a white precipitate of mercuric amido chloride, $Hg(NH_2)Cl$.

$$HgO_2 + 2NH_3 \rightarrow Hg(NH_2)Cl + NH_4Cl$$

It is used as an antiseptic due to its poisonous nature. It is also used as an intensifier for photographic negatives.

Mercury (II) Iodide (Mercuric Iodide, HgI_2): A scarlet precipitate of mercuric iodide is obtained when a calculated amount of potassium iodide is added to a solution of mercuric chloride.

$$HgCl_2 + 2KI \rightarrow Hg1_2 + 2KCl$$

It is used in the preparation of Nessler's reagent and in the form of ointment for the treatment of skin infection.

Mercury (1) Nitrate or Mercurous Nitrate, [$Hg_2(NO_3)_2$]: Prepared by heating mercury with excess of conc. nitric acid and keeping the solution over sulphuric acid for sometime when deliquescent crystals of mercuric nitrate, $Hg(NO_3)_2.2HgO$ are obtained.

Mercury (1) Oxide (Mercurous Oxide, Hg_2O): A black amorphous powder. In light it slowly decomposes to form mercury (II) oxide

$$Hg_2O \rightarrow HgO + Hg$$

It is formed as a black precipitate adding sodium hydroxide to the solution of a mercury (I) salt in cold

$$Hg_2(NO_3)_2 + 2NaOH \rightarrow Hg_2 + 2NaNO_3 + H_2O$$

Mercury (II) Oxide (Mercuric Oxide, HgO): A yellow solid is obtained by adding a solution of caustic soda to the solution of mercuric chloride

$$HgCl_2 + 2NaOH \rightarrow HgO\downarrow + 2NaCl + H_2O$$

On heating mercury to its boiling point in air or oxygen or by carefully heating mercuric nitrate forms a red mass

$$2Hg(NO_3) \rightarrow \underset{\text{Red Mass}}{2HgO} + 4NO_2 + O_2$$

It exists in two forms, yellow and red.

The difference in colour is due to the difference in sizes of the two forms.

It is insoluble in water and dissolves in hot concentrated hydrochloric acid. It is poisonous and is used for making yellow ointment for sore eyes. It is also used as pigment in paints.

Mercury (I) Sulphate (Hg_2SO_4): A white crystalline solid prepared by adding a soluble sulphate to a solution of mercury (I) nitrate

$$Hg_2(NO_3)_2 + Na_2SO_4 \rightarrow Hg_2SO_4 + 2NaNO_3$$

It is sparingly soluble in water. It Is used in the weston standard cell.

It is also used to catalyse the oxidation of organic substances.

Mercury (II) Sulphate (Hg_2SO_4): A white opaque mass obtained by heating mercury with excess of sulphuric acid

$$Hg_2 + 2H2SO_4 ® HgSO_4 + SO_2 + 2H_2O$$

It is hydrolysed by water to give red basic. sulphate, $HgSO_4.2H_2O$. It is used as catalyst for the conversion of acetylene into acetaldehyde.

Mercury (II) Sulphide (HgS): A black precipitate obtained by passing hydrogen sulphide through a solution of mercuric chloride. It occurs in the nature as cinnabar. It is insoluble in water. On heating, it changes into mercury and sulphur which burns in air to give sulphur dioxide. It exists in two forms red and black. The red variety is used as pigment under the name vermilion.

Mercury Vapour Lamp: A lamp emitting a strong bluish light by the passage of an electric current through mercury vapour in a bulb. The light is rich in ultraviolet radiations; used in artificial sunray treatment and in street lighting.

Mesityl Oxide ($(CH_3)_2C:CHCOCH_3$: A colourless oily liquid, b.p. 130°C, used as a solvent and in organic synthesis.

Meson: A particle which has the charge of the electron, either positive or negative and a mass approximately equal 100 times that of electrons.

Metal: The elements that are good conductor of electricity and whose conductivity decreases as the temperature rises. So the metals have small attraction for electrons. Thus the metallic character increases from top to bottom in a group and decreases from left to right in a period in a periodic table. The metals are generally melleable and ductile.

Metal Bond: A metal is defined as a substance consisting of positively arged charged ions, fixed in the crystal lattice with negatively charged electrons moving freely through the crystal. The free electrons act as a cohesive force, without which the positive ions alone would repel each other.

Metaldehyde: A white, inflammable, poisonous solid polymer of acetaldehyde. Used as fuel in small heaters.

Metallic Crystals: The type of crystal formed by most metals, in which the outer electrons of the metallic atoms are shared by the crystal as a whole. Thus, the positively charged metal ions in the crystal lattice are surrounded by a 'gas' of free electrons. These free electrons account for the fact that most metals are good conductors of heal, and electricity.

Metallic Soap: An insoluble salt formed by a metal and a fatty; acid (especially salts of lead and aluminium). Used for water-proofing textiles and as a drier for paints.

Metalloids: The elements resembling nonmetal as well as metals in their properties, e.g., boron, silicon, arsenic and antimony.

Metamerism: A type of isomerism exhibited by organic compounds of the same chemical class or type.

Metamorphism: The transformation of the structure or constitution of rocks due to heat and pressure.

Methanal (Formaldehyde, HCHO): A colourless, pungent smelling gas, b.p. 21°C. It is very soluble in water. It is used as disinfectant, antiseptic and to preserve biological specimen. It is also used to manufacture dyestuffs, and formaldehyde resins (bakelite).

Methane (CH_4): A colourless, odourless, nonpoisonous gas; the b.p. is 164°C and m.p.-184°C. It occurs in natural gas and the gases from the oil wells, It is the first compound of alkane homologous series. It reacts with Cl_2 in the presence of' sunlight. It is used to produce hydrogen gas, methyl chloride, methylene chlorine, methanol and formaldehyde.

Methanamide or Formamide ($HCONH_2$): A hygroscopic liquid readily soluble in water and alcohol. It is manufactured by the catalytic combination of ammonia and carbon monoxide under high pressure.

$$CO + HNH_2 \rightarrow HCONH_2 + H_2O$$

It is used as a solvent and as a plasticizer.

Methanamine or Methyl Amine (CH_3NH_2): A colourless, inflammable liquid (b.p. 292 K) with a fishy ammoniacal odour. It is obtained industrially by heating ammonium chloride with two equivalents of formaldehyde. It is used as refrigerant.

Methanoic Acid (Formic Acid, HCOOH): A pungent, corrosive, liquid, m.p. 8.4°C, b.p. 100.5°C. It is miscible with water, ethanol, and ether in proportion. It is used as an antiseptic and for preservation of fruit. It is also used in dyeing wool and cotton fibre.

Methanol (Methyl Alcohol, Wood Alcohol, Carbinol CH_3OH): A colourless inflammable liquid b.p. 64°C and is poisonous. It is used as a solvent for varnishes, paints, celluloid, cement and shellac. It is used in the manufacture of dyes, perfumes and formaldehyde. It is also used to prepare methylated spirit and automobile and freeze mixtures.

Methylated Spirit: Ethanol to which is added methanol (about 9%), pyridine (about 0.5%) and blue dye.

Methyl Benzene (Toluene, $C_6H_5.CH_3$): A colourless liquid b.p. 111°C resembling benzene in many of its chemical properties. It is used in the manufacture of explosives (e.g., T.N.T.), drugs, dyestuffs, saccharine and printing ink.

Methyl-cylohexanol ($CH_3C_6H_{10}OH$): A colourless viscous liquid consisting of a mixture of isorners with b.p. in the range 167–174°C, it is from cresol and used as a solvent for rubber and cellulose.

Methylene Blue ($C_{16}H_{18}N_{13}SCl$): A soluble, intense blue dye. Used as dyestuff, in medicine and as a stain in biology.

Methyl Orange: An acid dye obtained by coupling diazotised sulphanilic with dimethylaniline. It is used to dye wool, silk and as an indicator in acid alkali titration

Methyl Red: An acid dye obtained by coupling diazotised oaminobenzoic acid with dimelhylaniline. It is used as an indicator in acidalkali titration.

Methyl Salicylate (O-HOC_6H_4 $COOCH_3$): A colourless pleasant-smelling liquid (b.p. 295 K) obtained by direct esterification by distilling salicylic, acid with methanol in the presence of hydrogen chloride.

Methyl Sulphate ($(CH_3)_2SO_4$): A colourless syrupy liquid, b.p. 461 K, immiscible with water. It is poisonous in nature. It is used in the laboratory for the introduction of methyl radical in alcohols, phenols and amities.

Methylated spirit: A liquid fuel and solvent containing rectified spirit (90%), methanol (9.5%) and pyridine (0.5%). used for dissolving and paints.

Metol, p-Methylaminophenol, $CH_3NH.C_6H_4OH$: A white crystalline compound m.p, 87°C, as a developer in photography.

Mica: A group of minerals, the most important of which are muscovite and phlogophite.

Micelle: The term used to show the colloidal particles in aggregation colloids, e.g., dyes and soaps.

Microcosmic Salt [Na (NH_4).HPO_4.$4H_2O$]: Prepared by dissolving molecular proportions of Na_2HPO_4 and NH_4Cl is hot water and crystallizing. It is being less soluble of all crystalizes out first. It is filtered washed with water to remove traces of sodium chloride and recrystallized.

$$Na_2HPO_4 + NH_4Cl \rightarrow Na(NH_4)\ HPO_4 + NaCl$$

It is used for detecting silica which being insoluble in $NaPO_3$ gives a clouy bead.

Mineral Colour: Inorganic pigments like prussian blue, chrome yellow, Iron buff etc. They are used for colouring paints paper and plastics..

Mineral Oil: An oil having mineral origin and mixture of hydrocarbons, e.g., kerosene.

Misch Metal: An alloy of cerium with small amounts of oilier rare earth metals. Used for 'flints' in automatic lighters.

Mitscherlich's Law: The substances that crystallize in isomorphous forms (i.e., have identical crystalline forms and form mixed crystals) have similar chemical compositions. The law can he used to indicate the formulae of the compounds. For example, die fact that chromium oxide is isomorphous with Fe_2O_3 and Al_2O_3, implies that its formula is Cr_2O_3.

Mixed Indicator: The use of mixture of two or more indicators in order to sharpen the colour change or to restrict to pH range.

Mixture: The system in which two or more homogeneous substances (elements or compounds) are simply mixed together in any proportion. All mixtures (except solutions) are heterogenous.

Moderator: A substance used in nuclear reactors to reduce the speed of fast neutrons produced by nuclear fission.

Molar Mass: The number of grams per mole of a compound or element.

Molar Volume: The volume occupied by one mole of molecules on any gas at standard conditions (STP), e.g., as STP 32 g of oxygen occupies 22.4 litres.

Molar Method: Titration of chloride ion with silver ion in the presence of added chromate ion, A red precipitate forms at the end point.

Molar Solubility: The number of moles of a solute required to give 1 liter of a saturated solution of the solute.

Molality (Symbol m): The number of moles of solute in 1000 gm solvent. It is useful unit of calculation dealing with the freezing and boiling point solutions.

Molarity (Symbol M): The number of moles of solute in 1 litre of solution. It is very convenient unit for laboratory work, since aqueous solutions of known molarity can be prepared easily by weighing out small amounts of solute and measuring the volume oil solution in calibrated containers.

Molecular Crystal: A crystal that has molecules or individual atoms at the lattice points.

Molecular Equation: A chemical equation giving the full formulas of all the reactants and products that will be involved in an actual experiments.

Molecular Formula: The formula of a compound telling the actual number of atoms of various elements in the molecule, i.e.,

Molecular formula = (Empirical formula)$_n$

where n is a whole number.

Molecularity: The total number of molecules involved in the chemical change, i.e., hydrolysis of ester.

$$CH_3COOH + H_2 \rightarrow CH_3COOH + CH_3OH$$

The molecularity of the above reaction is two as 2 molecules take part in the reaction.

Molecule: The smallest part of a chemical compound, formed by the combination of atoms in a whole number ratio, that takes part in a chemical reaction, e.g., HCI is a molecule of hydrogen chloride. Many molecules of natural products and polymers are so large that they may contain thousands of atoms. Such molecules are called macromolecules, e.g., a protein molecule.

Molecular Orbital: When two atomic orbitals overlap they coalesce to produce a single molecular orbital.

Mole Fraction: The number of moles of a given component in a mixture divided by the total number of moles present of all the components.

The mole fraction of compound A is

$$nA/(nA + nB + nC + ...)$$

where nA is the number of moles of A etc.

Molisch's Test: Add a few drops of alcoholic solution of naphthol to a solution of glucose. Slip about. 1 cm^3 of concentrated sulphuric acid along the sides of the testtube, when a violet colour is obtained. This test is given by all Carbohydrates.

Molybdenum: A silvery white metal which is fairly hard. It exists in nature as molybdenite, MoS_2 and Wulfenite, $PbMoO_4$. Its compound are used as blue pigments for printing and dyeing wool.

Symbol : Mo; a.n. 42, m.p. 2620°C; bp. 4800°C; rd, 101.

Molybdenum (II) Chloride ($MoCl_2$): A pale yellow solid obtained by direct combination of elements. It is not attacked by water. It is stable in air, It is soluble in some organic solvents like alcohol, ether, acetone and pyridinc. Blue Molybdenum oxide (MO_3O_8): A dark blue substance obtained by reducing molybdic acid with zinc or sulphur dioxide.

Molybdenum (III) Chloride ($MoCl_3$): A red amorphous powder obtained by reducing pentachloride with hydrogen, It is insoluble in water which, however decomposes it on boiling. On heating $MOCl_3$ disproportionates with $MoCl_2$, and $MoCl_4$.

Molybdenum (IV) Chloride ($MoCl_3$): A brown solid obtained by heating MoO_2 with a solution of chloride in CCl_4 at 520 K. On heating strongly it decomposes into tri and pentachlorides.

Molybdenum (V1) Fluoride (MoF_6): A white solid formed by the union of the elements. It is extremely reactive and hydrolysed by water Molybdenum (V) Chloride ($MoCl_5$): A dark green solid obtained by direct combination of elements . It reacts with water with a hissing sound forming $MoOCl_3$.

Molybdenum Acid Trioxide (Molybdic Anhydride, MoO_3): A yellow crystalline substance that loses a molecule of water at 70°C to form molybdenum trioxide (molybdic anhydride).

Monazite: A mineral containing phosphates of cerium, thorium, and other rare earths, with some occluded helium.

Mond Process: The extraction of nickel by the action of carbon monoxide CO, on the impure metal, This gives nickel carbonyl $Ni(CO)_4$ a gas that decomposes when heated to 200°C into pure nickel and carbon monoxide, the latter being used again. Named after Ludwig Mond.

Monel Metal: An alloy of copper (255-355), nickel (605-705) and small amounts of iron, manganese, silicon and carbon. Used as an acid-resisting material in chemical industry.

Monoclinic: Relating to crystals which have three unequal axes with one oblique intersection.

Monosaccharide: A sugar which cannot be hydrolysed to simple carbohydrates of smaller carbon content, e.g., glucose and fructose.

Morphine ($C_{17}H_{19}O_3N$): The chief alkaloid present in opium obtained from poppy. Morphine hydrochloride is commonly used in medicine as a soporfic, arid for alleviation of pain.

Morpholine: A colourless hygroscopic liquid used as a solvent for resins and waxes.

Mosaic Gold: Crystalline stannic sulphide consisting of shining, golden yellow scales.

Mother Liquor: This solution left after the formation of crystals.

Mould: A vessel into which a hot liquid material is poured to become cold. When cold it solidifies in the shape of the mould.

Muffle Furnaces: A furnace designed so that the charge does not come into contact with the hot combination gases.

Multicentre Bond: A bond obtained by a bonding molecular orbital between three or more atoms containing a single pair or electrons. This type bonds are common in boron hydrides.

Multiple Bond: A bond between two atoms involving more than one pair of electrons, e.g., double bond in oxygen and triple bond in nitrogen.

Multi-Proportions Law: Two elements A and B combines to form more than one compound, then the different amounts of A that combine with a fixed amount of B are always in a simple multiple ratio. The ratio of oxygen in CO and CO_2 is 1:2.

Mumetal: An alloy of high magnetic permeability containing up to 78% nickel in addition to iron, copper and manganese.

Muscone or 3-methyl-cyclopentecanone ($C_{16}H_{30}O$): A yellow liquid (b.p.330°C). The perfume was obtained from the scent glands of the Tibetan musk deer.

N

Naphtha: A mixture of hydrocarbons in various propordons, obtained from paraffin oil, coaltar, etc. Wood naphtha is impure methanol, CH_3OH, produced by the destructive distillation of wood.

Naphthalene ($C_{10}H_8$): Colourless lustrous plates. It is insoluble in water but soluble in hot ethanol, cold ether, benzene etc. Chemically naphthalene resembles benzene in many of its reaction. It is used as insecticides, and in the preparation of phthalic anhydride and dyes.

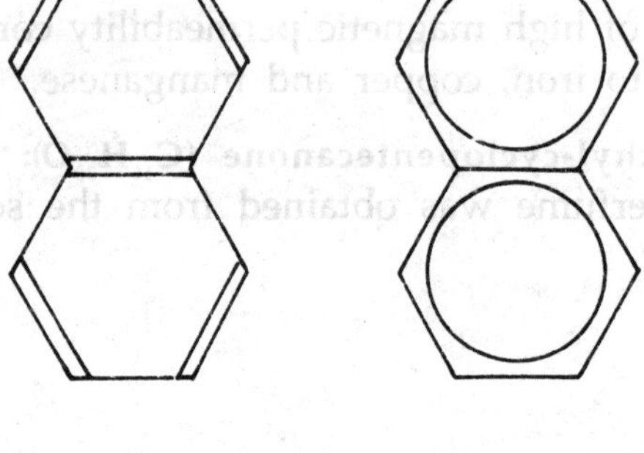

Representations of naphthalene

1, 2-Naphthaquinone: A nonvolatile odourless, red needle obtained by oxidation of 1-amino-2-naphthol with dichromate and sulphuric acid.

1, 4-Naphthaquinone: A volatile yellow solid obtained by oxidation of 1, 4-naphthalene diamine, 1, 4-naphthalene diol or 4-amino-1-naphthol with acid dichromate.

2, 6-Naphthaquinone: A nonvolatile, odourless orange solid obtained by oxidation of 2, 6-naphthalenediol dissolved in benzene with lead dioxide.

Naphthenes: Cycloalkanes; saturated hydrocarbons having at least one closed ring of carbon atoms.

Naphthol ($C_{10}H_7OH$): Two isomeric derivatives of naphthalene both of which darken in colour on exposure to light. Alphanaphthol is used in the manufacture of and perfurnes. Betanaphthol is used as an antiseptic and in the manufacture of dyes, drugs and perfumes.

1-Naphthylamine (a-Naphthylamine, $C_{10}H_9N$): A colourless crystal. It is prepared by the reduction of 1-nitronaphthalene with iron and a trace of hydrochloric acid.

2-Naphthylamine (b-Naphthylamine, $C_{10}H_9N$): A lustrous crystal obtained from 2-naphthol by heating it with a strong solution and ammonia at 150°C. It used as an endonent in a few azodyes.

Naphthoyl: The univalent radical $C_{10}H_7CO$– (from naphthoic acid, $C_{10}H_7$ COOH).

Narecine ($C_{23}H_{27}NO_{83}H_2O$): A hydrogen just produced by the electrolysis of water or from acids by the action of metals is in the atomic state and possesses enormous intrinsic energy. The reducing power is very must as compared to molecular gaseous hydrogen. It is used to produce hydrides of antimony, arsenic and phosphorus.

Nascent Hydrogen (New Born): A hydrogen just produced by the electrolysis of water or from acids by the action of metals is in the atomic state and possesses enormous intrinsic, energy. The reducing power is very much as compared to molecular gaseous hydrogen It is used to produce hydrides of antimony, arsenic and phosphorus.

Nascent State: Certain elements, notably hydrogen, are more active when being set free in a chemical reaction that in their ordinary state n such as nascent' elements are though to owe their activity to being composed of single atoms instead of molecules, or alternatively to some of the chemical energy liberated on the reaction being associated with the hydrogen instead of being released in the form of heat.

Natron: Natural sodium sesquicarbonale, Na_2CO_3 $NaHCO_3.2H_2O$.

Natta Process: A procedure for the manufacture of isotactic polypropylene using Ziegler catalysts

Natural Abundance: The abundance of each different isotope in an element as it is normally found in nature.

Natural Gas: A mixture of gaseous hydrocarbons, predominantly methane, often containing other gases, issuing from the Earth in some localities, more particularly near deposits of mineral oil.

Natural Rubber: Obtained from latex which is coagulated with aceticacid or formic acid and then subjected to treatment like squeezing, rolling, milling and Vulcanizing. It is tough, elastic, solid Pure rubber is transparent, amorphous solid which on streching or prolonged cooling becomes crystalline. It is insoluble in water alcohol, acetone dilute acids and soluble in ether.

N, 1-Dimethylaniline [$C_6H_5N\ (CH_3)_2$]: An oily liquid. It is obtained by heating and aniline with methanol (in excess) and sulphuricacid at K under pressure. It is used as a dye intermediate and for obtaining dimethyl amine.

Neel Point: The temperature at which magnetic susceptibility becomes normal.

Negative Adsorption: When adsorption at a surface takes place from a solution, the concentration of some components may be higher at the surface than in the bulk, whilst with other components the concentration will necessarily be less at the surface than in bulk. These latter components are said to undergo negative adsorption.

1-Naphthylamine (a-Naphthylamine, $C_{10}H_9N$): A colourless crystals, m.p. 50°C. It is prepared by the reduction of 1-nitronaphthalene with iron and a trace of hydrochloric acid.

2-Naphthylamine (B-Naphthylamine, $C_{10}H_9N$): A lustrous crystals (m.p. 112°C) obtained 2-naphthol by heating.

Nematic Liquid Crystal: Composed of long, rod like molecules packed like short pieces of uncooked spaghetti.

Neodymium: A silvery element having two allotropic forms and belonging to the lanthanoid series of metals. It is a toxic element that occurs in association with other lanthanoids. Used in various alloys, as a catalyst, and in compound in carbonarc searchlights, and in the glass industry. Symbol Nd; m.p. 1010°C, b.p. 3127°C.

Neon: A monoatomic, colourless, inert gas. It exists in air in small amount It is used in neon signs and lights.

Symbol Ne; a.n. 10; m. p 248.67°C, b.p. 246.05°C, r.d. 0.9 kg m^3

Neoprene: A type of synthetic rubber made by polymerization of 2-chlorobuta-1, 2 diene. It is more resistant to oil, solvents and temperature natural rubbers.

Nephelometry: A method of quantitative analysis which involves the spectrophotometric estimation of the scattering of light by colloidal suspension of a precipitate.

Nephelometry: A radioactive Silvery element of the actinoid series of metals. Found on earth only in minute quantities in uranium ores. Obtained as a by-product in 239 Pu production.

Symbol Np; m.p. 640°C, b.p. 3902°C.

Nerol, $C_{10}H_{17}OH$: A colourless liquid unsaturated alcohol, isomeric with geraniol, used in perfumes and obtained from neroli oil.

Nerolidol ($C_{15}H_{26}O$): A sesquiterpene alcohol on oxidation with chromicacid it gives furnesal.

Neroli Oil: An essential oil obtained from flavours of orange trees.

Nerve Gas: A war gas which attacks the nervous system especially the nerves controlling respiration. Most nerve gases are oil derivatives of phosphoric acid.

Nervone ($C_{48}H_{91}NO_8$): A cerebroside, crystallizing in needles from alcohol (m.p. 180°C). It hydrolyses to D-galactose, sphingosine and nervonic acid.

Nervonic Acid or Cis-14-Tetracosanic Acid ($C_{24}H_{46}O_2$): A crystalline (m.p. 43°C) formed by hydrolysis of nervone.

Nessler's Solution: A solution of potassium mercurie iodide, $KHgI_3$, in potassium hydroxide solution. Used at a test for ammonia, with which it forms a brown colouration or precipitate.

Nessler Tubes: Cylinder of thin, mainly graduated used for comparing turbidities and colours of solutions.

Neuraminic Acid: An important amino sugar which is widely distributed in animal tissues.

Neurine or Trimethyl Vinyl Ammonium Hydroxide or Trimethylene($C_5H_{13}NO$): A liquid forming crystalline trihydrate. It is formed as a product of putrefaction of lecithin. It is obtained synthetically from choline and decomposes easily to, trimethylamins, It is found free and combined in brain.

Neutral: Neither acid nor alkaline. Containing equal numbers of hydroxyl and hydrogen ions and having a p^H of 7.

Neutralization: Reaction of an acid and a base in volumetric analysis. The neutralization point or end point is detected with indicators.

Neutrino: A stable particle with no electric charge or rest mass, but with spin 1/2.

Neutron: One of the fundamental particles in an atom. It has same mass as a proton but is electrically neutral.

Neutron Activation Analysis: A technique for analyzing for trace impurities by rendering them radioactive through neutrin bombardment and then studying the frequencies and intensities of the gamma radiations they emit.

Neutron Flux: A measure of the number of neutrons passing through 1 sq em in any direction in 1 second.

Neutron's Law of Cooling: The rate at which a body loses heat to its surroundings is proportional to the temperature difference between the body and its surroundings.

Neutron Number: The number of neutrons in the nucleus of an atom.

Neutron Temperature: The energies preserved by neutrons in thermal equilibrium with their surroundings may be expressed in terms of a temperature if it is assumed that they behave as a monoatomic gas. Under these conditions the neutron temperature on the Kelvin scale, T, is given by $E = 3KT/2$, where E is the neutron energy and K is Boltzman's constant.

Newland's Law of Octave: In 1865, Newland, an English chemist, suggested that if the elements are arranged in order of increasing atomic mass, a similar set of properties is displayed by every eighth element.

Newton (Symbol N): The SI unit of force, equal to the force required to accelerate one kilogram by one metre second^{-2}.

$$1\ N = 1\ kg\ ms^{-2}$$

N, N-Dimethylaniline ($C_6H_5N(CH_3)_2$: An oily liquid (b.p. 466 K), It is obtained by heating aniline with methanol (in excess) and sulphuricacid at 500 K under pressure, It is used as a dye intermediate and for obtaining dimethyl amine.

Nachrome: Trade name for a nickel-Chrormium alloy used for wire in electrical devices, owing to its high resistance and its ability, to with stand high temperatures.

Nickel: A silver white metal which can take very good polish. It occurs in combined form, nickelite Ni-As, nickel blends NiS, and nickel glance, Ni-As S. It is very ductile and car be easily forged, welded and rolled in to sheets. It is resistant to air and moisture. It is used in making alloys, in electroplating, and in making nickel steel and stainless steel. Symbol Ni; a.n. 28; m.p. 1400°C; b.p. 2840°C; r.a.m. 58-70.

Nickel Acetate [$(CH_3COO)_2N.4H_2O$]: A green crystalline soluble substance, used in nickel plating.

Nickel (II) Carbonate or Nickel Carbonate ($NiCO_3.6H_2O$): A green crystalline solid obtained by adding sodium bicarbonate solution to a nickel salt solution. However a basic carbonate of nickel, $NiCO_3$.´ $Ni(OH)_2$ is obtained by adding washing soda (Na_2CO_3) to a nickel solution. The normal carbonate is insoluble in water but soluble in acids. On beating without contact of air it decomposes into nickelous oxide. It forms a starting substance for the preparation of nickel catalyst. It is used in ceramic and electroplating.

Nickel Amines: Anhydrous nickel salts and nickel salts in the presence of aqueous ammonia amines, e.g. $[Ni(NH_3)_6]^{2+}$.

Nickel Ammonium Sulphate [$Ni(NH_4)2(SO_4)_2.6H_2O$]: A blue green crystalline substance obtained from a solution of the components. It is used in electroplating.

Nickel (II) Chloride or Nickel Chloride ($NiCl_2.6H_2O$): A green crystalline hexahydrate obtained by dissolving the metal in aqua regia or by the action of dilute hydrochloric acid on nickclous oxide or carbonate.

Nickel Carbonyl [$Ni(CO)_4$]: A colourless liquid obtained by passing carbon monoxide over finely divided nickel at a temperature of approximately 60°C. It is used as catalyst.

Nickel (II) Chloride or Nickel Chloride ($NiCl_2.6H_2O$): A green crystalline hexahydrate obtained by dissolving the metal in aqua regia or by the action of dilute hydrchloric acid on nickclous oxide or carbonate. It loses its water of crystallizadon on beating in the absence of oxygen. It forms double salt with ammonium chloride ($NiCl_2$. $NH_4Cl.6H_2O$).

Nickel (II) Hydroxide or Nickelous Hydroxide $Ni(OH)_2$: A green precipitate obtained on treating a nickelous salt solution with caustic soda. The precipitate dissolves in excess of ammonia forming blue solution which contains the complex ion, $Hi(M_3)_6^{2+}$.

Nickel (II) Nitrate or Nickel Nitrate [$Ni(NO_3)2.3H_2O$]: Formed by dissolving the metal oxide or carbonate in dilute nitric acid and crystallizing product. It forms very soluble green hydrate $Ni(NO_3)_2.6H_2O$.

Nickel (II) Nitrate or Nickel Nitrate [$Ni(NO_3)2.3H_2O$]: Formed by dissolving the metal oxide or carbonate in dilute nitric acid and crystallizing product. It forms very soluble green hydrate $Ni(NO_3)_2.6H_2O$.

Nickel (II) Oxide or Nickelous Oxide (NiO): A greenish powder obtained by heating nickel hydroxide or carbonate out of contact with air. It is a basic oxide insoluble in water but soluble in acids. It is reduced to metal by heating it in hydrogen.

$$NiO + H_2 \rightarrow Ni + H_2O$$

Nickel (III) Oxide or Nickelite Oxide (Ni_2O_3): A black powder obtained by heating nickelous (NiO_2) oxide in air or by heating nickel nitrate cautiosly to about 300°C in air. It is insoluble in water. It dissolves in concentrated hydrochloric acid evolving chlorine and in oxy acids giving out oxygen.

Nickel (IV) Oxide or Nickel Dioxide (NiO_2): A black monohydrate($NiO_2.H_2O$) obtained by passing chlorine through a suspension of nickel hydroxide in water or by adding sodium hypochlorite or hypobromite to a nickel salt solution in the presence of an alkali.

$$NiCl_2 + NaOCl + 2NaOH \rightarrow NiO_2.H_2O + 3NaCi$$

It resembles manganese dioxide in chemical properties.

Nickel Plating: Depositing a thin layer of metallic nickel by an electrolytic, process.

Nickel Silver: A group of alloys of copper, zinc and nickel in ranging proportions.

Nickel Steel: Steel containing up to 6% nickel.

Nickel (II) Sulphate or Sulphate: A green crystal of hydrale $NiSO_2.7H_2O$ obwined by dissolving the carbonate indilute sulphuric acid, At high temperature two polymorphic forms of NiSO4.6H2O are obtained, one green and the other blue. It forms a bluish green, double salt, $(NH_2)_2SO_4$ $.NiSO_4.6H_2O$ which resembles mohor's salt. It is used in varnishes, on paints as moderates in dyeing and in nickel plating.

Nicotinamide C5H4 NCONH2: Vitamin of the Bcomplex. A colourless soluble substance used in medicine to treat pellagra.

Nicotine ($C_{10}H_{14}N_2$): A colourless oily liquid (b.p. 519.2 K under 730 mm pressure) which is readily soluble in water. Pure nicotine has an unpleasant smell unlike that of tobacco.

Nicotinc Acid, $C_5H_4N.COOH$: Vitamin of the B-complex. A colourless crystallinc solid found in yeast and meat.

Ninhydrin: A colourless organic compound that gives a blue colouration with amino acids. It is used as a test for amino acids.

Niobium: A transition element used in welding, special steels and nuclear reactors.

Symbol Nb; m.p. 2468°C; b.p. 4742°C.

Niobium Oxides (Nb_2O_5): A dense white material obtained by dehydration of the hydrated oxide formed by the hydrolysis of halides.

Nitramines: Nitro derivatives of amines with the general formula $R.NH.NO_2$

Nitrate: A salt or ester of nitric acid.

Nitrating Acid: A mixture of nitric and sulphuric acids. It is used for nitration.

Nitration: Introduction of a nitro group into an organic molecule. The nitration of aromatic compounds is an electrophilic: substitution, generally carried out by nitration with fuming nitric acid, or nitrationmixture ($H_2SO_4 + HNO_3$), die reaction being initiated by attack of anitroniurn ion, NO_2^+.

Nitrene Group (RN^+): The amino group nitrene derivatives, $(PR'_3)2NR$ and (WF_5NMe) are obtained by condensation reactions.

Nitric Acid (HNO_2): A colourless fuming liquid with a sharp chokingodour. It freezes at −42°C and boils at 80°C. it is miscible with water in all proportion. It reacts with many organic compounds to add —NO_2 or NO_3 groups to their molecules. It is used to make explosives, dyes, drugs, plastics and lacquers. Nitrates, produced by the action of the acid upon metals, oxides, or hydroxides, are valuable fertilizes and industrial chemicals.

Nitrides: Compounds of trivalent nitrogen and metals formed by passing nitrogen over red-hot metal.

Nitriding: A method of surface hardening steel by treating it in an atmosphere of NH_3 which gives a surface layer of nitride.

Nitrification: (1) The treatment of a substance with nitric acid. (2) The process of conversion, by the action of bacteria, of nitrogen compounds from animal and plant waste and decay, into nitrates in the soil.

Nitrile (Cyanide): A kind of organic compound containing the group. They are generally liquids with pleasant smell.

Nitrile Rubber: A copolymer of butadiene with acronitrile. It is used in sealing rings, gaskets and hoses.

Nitroanilines: O-and p-nitroaniline obtained by nitration of acetanilide followed by hydrolysis.

n-Nitroaniline is obtained by partial reduction of m-dinitrobenzene with ammonium hydrogen sulphide or sodium sulphide.

All the three isomeric nitroanilines are solid, sparingly soluble in water but readily soluble in alcohol. They are all used in the manufacture of dyes.

2-Nitroanisole or O-Nitroanisole ($C_7H_7NO_3$): A colourless oil obtained by heating 2nitrobenzene with NaOMe or from 2 nitrophenol and MeCl. It is an important as a dyestuffs intermediate after reduction to O-anisidine with Fe and HCl.

Nitrobenzene ($C_6H_5NO_2$): A pale yellow of, b.p. 211°C with smell like benzaldehyde. Its vapour is poisonous. It is obtained by the nitration of benzene. It is used as solvent, oxidising agent and in the manufacture of azodyes, aniline, floor polishes etc.

Nitrochalk: A mixture of calcium carbonate, $CaCo_3$, and ammonium nitrate, NH_4NO_3. used as a fertilizer.

Nitro-Compound: Compound containing NO_2^- group in its molecular structure.

Nitrogen: A colourless gas without any taste or odour found in the atmosphere as free nitrogen. It occurs in the combined state also as $NaNO_3$ and KNO_3 (salt petre). It is the first element of VA group of the periodic table. It is used in the manufacture of ammonia, nitric acid and nitroides. It has two 14 N, the common isotopes and 15 N about 0.3% abundant.

Nitrogen Cycle: Nitrogen is an essential constituent of animal and vegetable matter and is indispensible for the growth. Ultimate source of nitrogen is atmospheric nitrogen but neither plants nor animals are capable of assimilating free nitrogen. The nitrogen passes from atmosphere to plants and animals, converted into useful products ammonia, nitric acid and calcium cyanamide and still its percentage in the atmosphere remains practically unchanged. This is due to the fact that combined nitrogen is constantly passing back to the atmosphere.

Nitrogen Dioxide (NO_2): A compound consisting of two forms, the monomer, NO_2 and the dimer, N_2O_4 (dinitrogen tetroxide). Below the b.p. (21.15°C) it consists mainly of N_2O_4 Which is colourless. Its degree of dissociation into NO_2, which is a brown gas, increases with increase of temperature, the vapour at 150°C is black. It is formed by reduction of nitric acid and by the action of heat on some nitrates.Used as oxidant and for nitration.

Nitrogen Fixation: A reaction in which atmospheric nitrogen is changed into nitrogen compound.

Nitrogen Monoxide (Nitric Oxide NO): A colourless gas obtained by the action of dilute nitric acid on copper turnings. It is used as reducing agent.

Nitrogen Pentaoxide Nitrogen (V) Oxide (N_2O_5): Obtained by distilling concentrated nitric acid with phosphorus pentaoxide at 300 K in a glass retort when the former is dehydrated to nitrogen pentaoxide.

$$4HNO_3+P_4O_{10}\rightarrow 4HPO_3+2N_2O_5.$$

A white colourless solid (m.p. 303 K). It is a powerful oxidizing agent.

Nitro Glycerine: A colourless poisonous oily liquid obtained by adding glycerol in a thin stream to a cold mixture of concentrated nitric and sulphuric acids.

Nitrogen Sesquioxide or Dinitrogen Trioxide or Nitrogen (III) Oxide (N_2O_3): Obtained by passing a mixture of nitric oxide and nitrogendioxide through a tube cooled below 253 K A blue liquid is formed.

Nitrogen Triiodide (NF_3): A black powder obtained from NH_3 gas over $KIBr_2$. Alcoholic solution of I_2 and NH_3 give a black solid NH_3NI_3 which is stable when moist but shock sensitive when dry.

Nitrogen Trichloride (NCl_3): Obtained by the action of chlorine or sodium hypochlorite on ammonium chloride.

$$NH_4C1+3Cl_2\rightarrow NCl_3+4HCl$$

It is extremely explosive yellow oil. It is hydrolysed by water and rapidly by alkali to give ammonia and oxy acid or oxysalt.

Nitrogen Trifluoride (NF_3): Obtained by electrolysis of fused ammonium hydrogen fluoride, $NH_4F.HF$ at 398 K. It is not hydrolysed.

1-Nitroguanidine: A white crystalline powder obtained from calcium carbide via calcium cyanamide, dicyanamide and guanidine nitrate which is changed to nitroguanidine by he action of conc. H_2SO_4. It is soluble in hot water and used in some moderropellants to make them cooler and flashless.

4-Nitroluene ($C_7H_7NO_2$): A colourless crystal. It is obtained by mononitrating toluene. It is reduced to p-toluidine with iron and hydrochloric acid.

Nitromethane ($CH_3.NO_2$): A colourless oily liquid, b.p. 100.8°C, used as a solvent and in organic synthesis.

Nitron ($C_{20}H_{16}N_4$): Used in ethanoic acid solution as a precipitating agents for nitrates, PF_6^-, ClO_4^-, WO^{2-}_4 also form precipitates.

Nitronium ion: An electrophile; the cation NO_2^+ generated by the action of sulphuric acid on nitric acid during nitration reaction.

2-Nitrophenol or p-Nitrophenol ($C_6H_5NO_3$): A bright yellow needle shaped solid obtained together with 4-nitrophenol by careful nitration of phenol. It is used in photography and in dyestuffs.

4-Nitrophenol or p-Nitrophenol ($C_6H_5NO_3$): A colourless needle shape solid reduction with iron results 4-aminophenol.

Nitrourea ($NH_2CONHNO_2$): A colourless crystalline solid. It is obtained by adding urea nitrate to concentrated sulphuric acid at 0°C. It is used in the preparation of semicarbazide.

Nitrous Acid (HNO_2): A weak acid with a dissociation constant of 6.0 ´ 10^6 at 25°C It is used in the manufacture of dyes.

N-Methyl Aniline or N-Methyl Phenylamine or N-Methul Benenediamine ($C_6H_5NH.CH_3$): A colourless liquid. It is a stronger base than aniline. It is obtained by heating aniline with methanol and sulphuric acid at 500 K under pressure.

Nobelium: A radioactive transuranic element of the actinoid series not found naturally on earth. Several very shortlived isotopes have been synthesized.

Noble Gases: (insert gases, rare gases, group 0 elements) The group comprised of helium (He), neon (Ne), argon (Ar), krypton (Kr), Xenon (Xe) and radon (Rn). All these elements we monoatomic gases. The properties of the noble gases reflect their very stable electronic configuration. The atoms have no tendency to combine with each other to form molecules. All the noble gases occur in atmosphere but radon is produced by the radioactive decay.

Noble Metals: Metals such as, silver, gold and platinum, that do not corrode or tarnish in air or water, and are not easily attacked by acids. From the chemical point of view, unreactive metals are low in the ceectromotive series.

Nonanoic Acid, $CH_3(CH_2)COOH$: A colourless oily liquid, used in the manufacture of lacquers and plastics.

Non-Aqueous Solution: A solution in which the solvent is not water but may be organic or inorganic, protonic or nonprotonic. Nonaqueous solvents may be selfionized.

Non-Electrolytes: The substances whose aqueous solution under ordinary condition do not conduct electricity. They include many water soluble organic compounds like sucrose, ethyl alcohol and glycerol.

Non-Ideal Solution: A solution is formed when the process of mixing its components is accompanied by the evolution or absorption of heat. It does not obey Raouit's law.

Non-Metals: Any element that does not conduct electricity. So the nonmetals have great attraction for electrons, So the nonmetallic character increases along a period from left to right where as it decreases in a group from top to bottom in a periodic table. Non-metals generally have the opposite characteristics as compound with metals but there are some exceptions.

Non-Polar Compound: A compound which does not have a permanentdipole moment.

Non-Stoichèiometric Compound: Compound in which the ratios of the numbers of atoms present do not follow simple integers, many transiiion metal oxides, silicates are non-stoicheiometric.

Normality (Symbol N): The number of gram equivalent weights of solute in one litre of solution.

Normal Solution: A solution that contains one gram equivalent weight litre of solution.

Nuclear Energy: Energy released during a nuclear reaction as the result of conversion of mass into energy as per mass energy equation. It is the source of energy in nuclear reactors and nuclear weapons.

Nuclear Equation: A shorthand expression of a nuclear reaction that is balanced when the sums of the atomic numbers on either side of the arrow are equal and the sums of the mass numbers are equal.

Nuclear Fission: The breaking of the nucleus of an element with a high atomic number (e.g., uranium) into two approximately equal parts. It can be spontaneous or caused by the bombardment of nucleus by neutron.

Nuclear Force: The attractive force that acts between nucleons when they are extremely close together. The nuclear force replaces the repulsive electromagnetic interaction between protons at such proximities and holds the nucleons together in the atomic nucleus. The precise, nature of the nuclear force is not known.

Nuclear Fusion: The forming of an atomic nucleus from the nuclei of low-atomic number, large amount of energy is released. The reaction takes place at very high temperature.

Nuclear Isomers: Atoms of an element of the same mass but processing different rates of radioactive decay.

Nuclear Magnetic Resonance Spectroscopy: The measurement of radio frequency required for resonance or precession of a nucleus oriented in a strong magneticfield. As the orbital electrons 'shield' the nucleus to a certain extent from the applied magnetic field, at a given frequency nuclei in different electronic or chemical environments will resonate at different values of the applied field as compared to a standard compound. This phenomenon is known as chemical shift and is generally reported in d (ppm) value or *t* value (10-d) compared from the position of tetramethyl silane (TMS), an internal standard being zero (or 10,t). The chemical shift (ppm) is also given by 10^6 ´ shift of a resonance signal from TMS in cycles per second divided by the operational frequency of the instrument. This phenomenon of chemical shift enables the scientists in deduction of structures of complex molecules.

Nuclear Reaction: Any reaction, which involves a change in the nucleus of an atom, as distinct from a chemical reaction which involves the orbital electrons.

Nuclear Reactor: Atomic pile. An assembly in which a nuclear fission chain reaction is maintained and controlled for the production of nuclear energy, radioactive isotopes or artificial elements.

Nuclear Transmutations: The changing of atoms of one element into those of another by suitable nuclear reactions.

Nucleic Acids: Polymers in plants and animals that are involved in the chemistry of heredity and whose molecules, when hydrolyzed give a sugar unit, a phorphation, and set of four or five nitrugas containing, basic hetercyclic compounds.

Nucleon Number (Mass Number; Symbol A): The number of nucleons (protons plus neutrons) in an atomic nucleus.

Nucleophile: An electron rich ion or molecule which takes place in an organic reaction.

Nucleophilic Addition: A kind of reaction in which addition of small molecules to an unsaturated organic substance takes place.

Nucleophilic Reagents: Reagents which react at centres of low electron density. Nucleophilic reagents behave as electron donors.

Nucleophilic Substitution: A reaction in which the substitution of an atom or group of atoms in an organic molecule is substituted by another atom or group, e.g., hydrolysis of ethyl chloride.

$$C_2H_5Cl + OH^- \rightarrow C_2H_5OH\ Cl^-$$

Nucleoside: A compound formed from a nitrogenous base and a pentose sugar. The phosphorylated derivative of a nucleoside is called a neucleotide.

Nucleus: The protons and neutrons reside together in a very small volume within the atom. The simplest nucleus is that of hydrogen atom H being simply one proton. The most massive naturally occurring nucleus is 238 U of 92 protons and 146 neutrons.

Nuclide: The term nuclide is applied in general way to a nucleus with a specified number of protons and neutrons.

O

Octagon: An eightsided polygon. The angle between the sides of a regular octagon is 135°.

Octahedron: A polyhedron having eight faces.

Octahydrate: A crystalline compound containing eight molecules of water of crystallization per molecule of compound.

Octane Number (of fuel): The percentage by volume of *iso*-octane (C_8H_{18};2, 2, 4-trimethylpentane) in a mixture of iso-octane and normal heptane (C_7H_{16}) which is equal to the fuel in knock characteristics under specified test conditions.

Octane, C_8H_{18}: A hydrocarbon of the paraffin series that exists in several isomeric forms. A colourless liquid, b.p. 126°C, r.d. 0.704.

Octane Rating (Octane Number) of Gasoline: The percentage of iso-octane in the mixture of iso-octane and n-heptane which matches the fuel in knocking.

Octant Rule: This relates the sign and the amplitude of cotton effect curves with the geometry and substitution of cyclohexanone derivatives and has been applied especially to steriods. It may also be extended to other cyclic ketones.

Octaves, Law of Newlands: found that when the elements were arranged in order of increasing atomic weights, the first element was similar to the eighth, the second to the nineth, the third to the tenth and so on. He termed this relationship the 'law of octaves'.

Octet: An outermost shell in atom which contains eight electrons. The noble gases are almost inert because they have complete octet.

Octet Rule: An atom tends to gain or lose electrons until its outershell consist of eight electron.

Oenanthic Acid or Heptanoic Acid ($C_7H_{14}O_2$): An oily liquid (m.p. –9°C; b.p. 115-116°C/11 mm) obtained by die oxidation of heptanal. It is found in some natural oils and waxes.

Oersted: A unit of magnetic field strength in the e.g.s. system. Symbol Oe. It is equal to $10^3/14/W\ Am^{-1}$.

Ohm (Symbol W): The SI unit of electrical resistance equal to a resistance that passes a current of one ampere when there is an electrical potential difference of one volt across it.

$$1\ \Omega = 1\ VA^{-1}$$

Oil: Naturally occurring hydrocarbons, mainly aliphatic, used as a fuel after refining and as a feedstock for the petrochemical industry. Calorific value C $4.4 \times 10^7\ J\ kg^{-1}$. World production 2×10^9 tonnes per year.

Also used to describe an oily substance obtained from natural products, e.g., oil of anise.

Oil Cake: A mass of oilseeds from which the oil has been expelled in a press (expellers) or extracted by a solvent (extractions); used as cattle food.

Oil of Camphor: Two fractions of the natural oil of camphor are used medicinally. (1) Light oil of camphor is an almost colourless fraction containing a small amount of camphor, about 30% of cineole and the remainder terpenes. (2) Dark oil of camphor consists very largely of safrole.

Oil of Caraway: Obtained by distillation from the ripe fruits of carum carvi. It contains about 60% of carvone, and is used to relieve flatulence, especially in infants.

Oil, Synthetic: Natural mineral oils are composed of various hydrocarbons. It is possible to make similar products artificially from coal, etc., by combining carbon or carbon monoxide, with hydrogen.

Oleic Acid, $C_{17}H_{33}COOH$: An unsaturated liquid organic acid which occurs in the form of glycerides in many fats and oils.

Oleum ($H_2S_2O_7$): A colourless fuming liquid obtained by dissolving sulphur (II) oxide in concentrated sulphuric acid. It results in sulphuric acid on diluting with water.

$$H_2SO_4 + SO_3 \rightarrow H_2S_2O_7 \text{ (oleum)}$$

$$H_2S_2O_7 + H_2O \rightarrow 2H_2SO_4$$

Oleyl Alcohol [$CH_3(CH_2)_7$ $CH:CH(CH_2)_7CH_2OH$]: An unsaturated liquid alcohol, b.p. 205°C, used in organic synthesis.

Oligomer: A polymer having comparatively few monomer units in the molecule.

Olivine (Mg,Fe)2SiO_4: A mineral silicate of magnesium and iron. The transparent form is used as a gem.

Omium Ion: An ion formed by addition of a proton (H^+) to a molecule. The hydromium ion (H_3O^+) and ammonium ion (NH_4^+) are examples.

Open-chain Compounds: Organic compounds not derived from ring compounds.

Open-hearth Process: A method of making steel by heating pig iron in an Open-hearth furnace by burning a mixture of gas and air. Oxygen is injected through pipes to oxidize impurities and burn off carbon and lime is used to form a slag

Optical Activity: The substance which can rotate the plane of polarised light, e.g., can sugar is an optically active compound. The substance which can rotate the plane of polarised light to left side is called levo rotatory (–) and on right side called dextro rotatory (+).

Optical Electrons: In general, only the outermost electrons in an atom are concerned in producing the emission or absorption of visible light. Such electrons are termed optical electrons.

Optical Exaltation: An abnormal increase in the refractivity of an organic compound, over and above the additive amount to be expected from the bonds present, due to the presence of a system of conjugated double bonds.

Optical Isomerism: A form of isomerism in which the isomers differ in their optical activity.

Optical Purity: The ratio of a substance's specific rotation to the specific rotation of the pure enantiomer.

Optical Rotation: Rotation of the plane of polarization of plane-polarized light by an optically active substance.

Orbital: A region around the nucleus of an atom in which there is a high probability of finding electrons. There are two such electrons in the *s*-orbital (spherical shape) at any energy level. Other electrons move about three axes at right angles to each other. Such are called p-electrons (dumbell shape) are in p-orbitals. These could be as many as six-p-electrons in orbitals beginning at the end of the second shell. The 3rd, 4th and 5th shell contain not only s and p electrons but also *d* and *f* electrons.

Orbital Angular Momentum, L: The total angular momentum of a set of electrons in an atom or ion. For filled and empty shell, L is zero.

Orbital Electron: Plantary electron An electron contained within an atom; it may be thought of as orbiting around the nucleus, in a manner analogous to the orbit of a planet around the Sun.

Orbital Moment: The component of the magnetic moment of an atom or ion resulting from the motion of the electron around the nucleus.

Orbital Splitting: The raising of the degeneracy of a series of orbitals of equivalent energy, e.g., by placing in a crystal field.

Orcinol, $CH_3CH_3(OH)2H_2O$: A white crystalline substance, m.p. 107-8°C, that reddens on exposure to air. Used in the analytical detection of carbohydrates.

Order of Reaction: The sum of the powers of the concentration terms in the experimentally determined rate equation for the reaction.

Ore: The minerals from which the metals can be economically extracted.

Organelles: Organised micro structures inside ells to which certain chemical processes are confined, e.g. mitochondria, chloroplast etc.

Organic Acid: An organic compound that is able to give up a proton to a base; i.e., one that contains one or in some carboxyl groups or in some cases hydroxyl groups (e.g., phenol).

Organic Base: A molecule or ion possessing a lone pair of electrons that can be used for coordination with a proton. The common organic compounds that fulfil this condition owe their basic character to an oxygen or nitrogen atom.

Organic Chemistry: The branch of chemistry which deals with the chemistry of hydrocarbon and their derivatives.

Organo-chlorines: A group of synthetic pesticides which contain carbon, chlorine and hydrogen and which are generally very persistent with long residual effect, e.g., DDT, HCH etc. insecticides.

Organometallic Compounds: A compound in which a metal is directly linked to carbon. The compounds are generally colourless mobile liquids having low boiling point.

Organo Thiocyanate: Haloaryl and haloalkyl thiocyanoacetate and thiocyanates used as insectofungicides, pesticide adjuvants etc. Some of them can synergize other insecticides, e.g., bornyl thiocyano acetate (thanite) has been used as synergist for pyrethrins in household sprays.

Orgel Diagrams: Simple graphs showing the relation between the energies of various electronic states and the crystal field splitting.

Orion Fibre: Made from poly acrylonitrile. It is insoluble in common solvents, The fibre has unusual dimension stability in the dry and wet states. It has excellent resistance to sunlight, outdoor exposure, microorganisms, acids and solvents. It can be made to look like silk or wool.

Orsat Apparatus: A portable apparatus for determining the amount of carbon dioxide, oxygen and carbon monoxide in flue or exhaust gases. A measured volume of the gas is successively passed through three tubes the first of which contains potassium hydroxide to absorb the CO_2, the second alkaline pyrogallol to absorb O_2, and the third cuprous chloride in hydrochloric acid to absorb the CO.

Ortho: (1) Prefix denoting right, straight, correct. (2) Denoting adjacency in position in a hexagonal ring of atoms, particularly the benzene ring. Abbreviated to o-as a prefix in naming a compound, e.g., o-dichloro-benzene.

Orthochromatic Film: A photographic film sensitive to green in addition to blue and violet light, thus giving a more accurate representation of colours in monochrome than ordinary film.

Orthoclase Felspar: Natural potassium aluminium silicate.

Osmic Acid, Osmium tetroxide, O_5O_4: A colourless crystalline solid; solutions used as a stain for fat globules in microscopy.

Osmiridium: A natural alloy of osmium and irridium, with smaller amount of platinum, rhodium, and ruthenium. Hard and resistant to corrosion, used for tipping pennibs.

Osmium: A transition metal found associated with platinum. It oxidizes readily and is used in pen nibs, pivots, and electrical contacts. Symbol: Os; m.p. 3045°C; b.p.5027°C.

Osmosis: A process in which a solvent is separated from a solution with the help of semipermeable membrane.

Osmotic Pressure: The pressure needed to stop osmosis from pure solvent into a solution. Mathematically, it can be expressed as

$$\pi = MRT$$

Ostwald's Dilution Law: A law relating the dissociation constant, and the degree of dissociation (or ionization), a, of a weak electrolyte of concentration c moles per litre. This law states that for a binary electrolyte

$$K = c\alpha^2/(1 - \alpha)$$

an equation that applies with a fair degree of accuracy to weak organic acids and bases. Named after Wilhelm Ostwald.

Ostwald Ripening: A process of crystal growth in which a mixture of coarse and fine crystals of a substance are left in contact with a solvent. This results in a growth of the large crystals and the ultimate disappearance of the fine crystals.

Ouaban, $C_{29}H_{44}O_{12}$: A white crystalline glycoside obtained from wood; used as a heart stimulant.

Oxalic Acid (Ethanedioic Acid, COOH.COOH): A colourless crystalline solid, m.p. 1.89.5°C. It is poisonous and soluble in water acid ethanol but insoluble in ether. It is used in the manufacture of ink and bleaching, straw. Its antimony salts are used as mordants in printing and dyeing. Potassium ferrous oxalate is used in photography as a developer.

Oxidant (Oxydising Agent): A substance which is capable of giving oxygen to bring about combustion, e.g., hydrogen peroxide or liquid oxygen in rocket fuels.

Oxidation: The process in which an atom or a molecule losses an electron, combines with oxygen or with another electronegative element. The compound that removes the electron or acquires the electron is called the oxidizing agent. In electrolysis this process occurs at the anode. The oxidation state of an atom is indicated by the number of electrons lost or effectively lost by the neutral atom. Oxidation number of negative ions is negative, e.g. in oxidation of zinc with oxygen to form zinc oxide, ZnO, the oxidation number of zinc is +2 and that of oxygen is –2. Oxidation and reduction are always simultaneous.

Oxidation Numbers: An electrical charge appears on the atom when calculated by arbitrary rules.

Oxide: Binary compound with oxygen, e.g. magnesium oxide, MgO.

Oxidizing Acids: An acid that acts as an oxidizing agent, e.g., nitric acid or sulphuric acid.

Oxime: A class of organic substance which contains the C.NOH group, obtained by the reaction of an aldehyde or ketone with hydroxyl amino (NH_2OH).

Oxo-Process: A method of manufacturing aldehydes by passing a mixture of carbon monoxide, hydrogen and alkanes over a cobalt catalyst at high pressures (100 atmosphere and 15°C). The aldehydes can subsequently be reduced to alcohols.

Oxy-Acetylene Burner: A device for obtaining a very high temperature flame for welding.

Oxygen (O_2): A colourless gas without any odour. It occurs in atmosphere as free oxygen. Water which covers more than three-fourth of the earth contains about 89 per cent by weight of combined oxygen. It also found as metallic oxide, nitrates, sulphate and silicates in the earth's crust. It combines mostly with other elements. It is essential for respiration and it is used in breathing apparatus. Liquid oxygen is used as a fuel in rockets and missiles. It is used in the manufacture of sulphuric acid and nitric acid.

Symbol O; a.n. 8; m.p. 29.4°C; b.p. 1.83°C; d. 1.429 kg m^{-3}

Oxyhemoglobin: An oxygen molecule reacts with iron atom of the hemoglobin to form a species.

Oxygen Carrier: A molecule which can react directly with molecular oxygen (to form a peroxide or superoxide derivative, generally a complex) and can subsequently use the oxygen in an oxidation reaction, generally of an organic substrate e.g., haemoglobin.

P

Paint: Liquid containing a coloured material in suspension. The application of the paint to a surface, and the evaporation or hardening of the liquid, covers the surface with die pigment in the form of a skin. The liquid generally consists of linseed oil, a 'thinner' of turpentine or other volatile liquid and a 'drier' to accelerate drying or hardening of the linseed oil.

Packing Fraction: The mass defect divided by the mass number. For the lightest and heaviest elements, the packing is positive, intermediate element have a negative packing fraction.

Palladium: A transition silvery white metal, malleable and ductile. It is used as a catalyst in hydrogenation reactions, in jewellery and used to prepare alloys,

Symbol Pd; a.n. 46; m.p. 1552°C; b.p. 3140°C; r.d. 120.

Palladium Black: A finely divided precipitate formed by reducing solution of Pd salts. It is used as hydrogenation catalyst.

Palladium (II) Chloride or Palladous Chloride (PdCI2): A dark red solid obtained by heating palladium in a current of chlorine at red heat. The solution on evaporation yields crystals of dihydrate, $PdCl_2.2H_2O$. It is reduced by sulphur dioxide or carbon monoxide to palladium.

It is soluble in hydrochloric acid. It reacts with ethylene to form the product $[C_2H_4.PdCI_2]$ which decomposes by water giving acetaldehyde.

$$[C_2H_4.PdCI_2] + 2H_2O \rightarrow 2CH_3CHO + 2Pd + 2HCl$$

with propene $CH_3CH{=}CH_2$, the product obtained is acetone. Both these reactions are of commercial importance.

Palladium (II) Hydrous Oxide ($PdO.nH_2O$): A yellow gelatinous precipitate obtained on adding alkali hydroxide to a solution of palladous.

Palmitic Acid, $C_{15}H_{31}COOH$: A waxlike fatty acid which occurs in the form of tripalmitin in palm oil and many natural fats.

Panchromatic Film: A photographic film sensitive to light of all colours including red, thus giving a more accurate representation of colours in monochrome than orthochromatic film.

Pantothenic Acid, $C_9H_{17}NO_5$: A white insoluble solid member of vitamin B complex.

Papaverine, $C_{20}H_{21}NO_4$: A white insoluble alkaloid obtained from opium. Used in medicine as an antispasmodic.

Paper Chromatography: A solute mixture dissolved in water is placed on a strip of chromatograph paper and a second solvent is allowed to travel along the strip. As a result of continuous and repeated partitioning the rate of migration of each component is different which is responsible for their separation on the paper. It is mainly used for the separate of amino acids, carbohydrate etc.

Paper Factor: (+) Juvabione, isolated from Pinacea wood used for making paper in the United States. It is a juvenile hormone which inhibits the metamorphosis of Pyrrhocoris apterus larvae to adult stage.

Para: (1) Prefix denoting beside, beyond; or wrong, irregular. (2) Denoting positions at opposite apexes in a hexagonal ring of atoms, particularly the benzene ring. Abbreviated to p as a prefix in naming a compound; e.g., p-dichlorobenzene (alternatively, 1, 4-dichlorobenzene).

Parachlor: The existence of intermolecular attractive forces in liquids does not permit the use of molecular volumes for comparisons of molecular dimensions. The parachor P, where g is the surface tension, M the molecular weight and D and d the densities of the liquid and vapour respectively, is an empirical measure of molecular volume and was widely used to determine the constitution of organic and inorganic compounds.

Paraffin Oil (Kerosene): A mixture of hydrocarbons obtained in the distillation of petroleum.

Paraffin Series: An homologous series of hydrocarbons having the general formula C_nH_{2n+2}.

Paraffin Wax: A solid mixture of hydrocarbons obtained from petroleum.

Parahydrogen: Hydrogen molecules in which the spins of the two constituent nuclei are antiparallel.

Paraldehyde $(CH_3CHO)_3$: A liquid polymer of acetaidehyde, b.p. 124°C. Used in medicine as a hypnotic.

Paramagnetism: The magnetic properties caused by impaired clectrons.

Para Rosaniline: A dye obtained by oxidation of a mixture of aniline (2-molecules) and p-toluidine (1 molecule) with nitroberizene or arsenic acid. It dyes wool and silk directly and cotton after mordanting with tannin.

Paraquat, or 1, 1-dimethyl-4, 4-bipyridylium and dimethyl sulphate or dichloride: Both salts are white crystalline solid (m.p. 175-180°C) obtained by quaternization of 4, 4'-bipyridye with dimethyl sulphate or methyl chloride. Paraquat is a contact herbicide.

Paris Green (Schweinfurt Green, Cupric Acetoarsenite): Used as a pigment, insecticide and a wood preservative.

Parke's Process: A process of refining lead from silver as impurity. A small amount of zinc is added to molten impure lead. Silver is much more soluble in zinc than lead.

Partial Ionic Character: The electrons of a covalent bond between atoms or groups with different clectro-negativides will be polarized towards the more electronegative constituent; the magnitude of this effect can be measured by the ionic character of the bond.

Partial Pressure: In a mixture of a gases the contribution that one component makes to the total pressure. It is the pressure that the gas would have if it alone were present in the same volume.

partition Coefficient: If a solute dissolves in two nonmiscible liquids, the partition coefficient is the ratio of the concentration in one liquid to the concentration in the other liquid.

Partition Law: Distribution Law In a heterogeneous system of two or more phases in equilibrium, the ratio of the activities (or less accurately the concentrations) of the same molecular species in the phases is a constant at constant temperature. This constant is termed as the partition coefficient.

Parylene Polymers: A series of polymers derive from di-p-xylenc, $(CH_2C_6H_4CH_2)_2$. Used for dielectric coatings in electronic equipment.

Pascal: The SI unit of pressure equal to a pressure of one neutron per square meter. Symbol Pa.

Paschen Lines: The lines represent transitions to the n = 3 level. These lines occur in the infrared region.

Passive: A metal that is unreactive because of the formation of a thin protective layer.

Pasteurization: The process of removal of some nonspore-producing organisms from milk, butter and beer by the use of low heat or radiation.

Pauli Exclusion Principle: W. Pauli postulated that no two electrons can have all the four quantum numbers the same. In essence, it means that in the same orbital of the atom, Pauli principle permits two electrons only under the condition that they have spins in opposite directions.

P-Block Elements: The elements of main groups IIIA, IVA, VA, VIA,VIIA and O. The electrons occupy plevels.

P-Deficiency Symptoms in Plants: Dwarfing, erect and narrow leaves and reddish or purple coloration of stalks and veins due to want of phosphorus, an important constituent of energy rich compounds like adenine diphosphate (ADP), adenine triphosphate (ATP), adenine monophosphate (AMP), nucleic acids (RNA and DNA), phospholipids, phytin etc. (at. phosphorous deficiency symptoms).

Pearl: A secretion consisting mainly of calcium carbonate, $CaCO_3$, produced by various molluscs.

Pearl Ash: Potassium carbonate, K_2CO_3, made from wood ashes.

Peat: An early stage in the formation of coal from vegetable matter. It is an accumulation of partly decomposed plant material, and is used as fuel.

Pencil Lead: A mixture of graphic with clay in various proportions, to give different degrees of hardness.

Penicillin: The penicillins are group of bactericidal antibiotics having basic ring structure of pencillanic.

Pentachlorophenol: A white insoluble derivative of phenol used as a fungicide.

Pentahydrate: A crystalline compound with five molecules of water of crystallization per molecule of compound.

Penta: Prefix denoting five or five folds e.g. pentahydrate means a compound that has five molecules of water of crystallization per molecule of the compound.

Pentane (C_5H_{12}): A straight chain alkane obtained by distillation of crude oil.

Pentanoic Acid: A colourless liquid carboxylic acid used in making perfumes.

Pentosan: The group of polysaccharides which contain pentose resides only e.g., xylans contain D-xylose and arabans only L-arabinose.

Pentose: A monosaccharide containing five carbon atoms and having the general formula $C_5H_{10}O_5$. The most important pentose is ribose, which is a essential constituent of the nucleic acids.

Peptide: A compound of two or more amino acids formed by condensation of the NH_2 group of one acid and the carboxyl group of another. The peptide linkage, NHCO results.

Peptones: Organic substances produced by the hydrolysis of proteins by the action of pepsin in the stomach. They are soluble in water, and are absorbed by the body.

Peptide: A compound formed by linkage of two or more aminoacids grouped together.

Per Acid: An acid that contains an extra oxygen than in normal, e.g., peracetic acid (CH_3CO_3H); perchloric acid ($HClO_4$) etc. The organic peracids are prepared by the action of hydrogen peroxide on carboxylic acids and are good oxidising agents. These are also used as catalysts in radical reactions (antiMarkownikoff addition).

Perchloro Ethylene: Misnomer for Tetrachlorethylene CO_2=CCl_2, a grain fumigant/fungicide.

Percussion Cap: A device used in firearms. It consists of a small copper cylinder containing mercuric fulminate or other violent explosive that will explode on being struck, thus, initiating the explosion of the main charge.

Perfect Gas (1deal Gas): A theoretical concept of a gas that would obey the gas laws exactly. Such a gas would consist of perfectly elastic molecules, the volume occupied by the actual molecules, and the forces of attraction between them, being zero or negligible.

Period: The horizontal rows in the conventional periodic table. There are seven periods.

Periodic Acids: A series of unstable acids formed by the addition of water, to the hypothetical iodine heptoxide, I_2O_7, e.g., HIO_4. They are known in the form of their salts, the periodates.

Periodic Table: A table of elements arranged in order of increasing atomic number to show similarities in chemical nature between elements.

Perion L or Nylon-6: A polyamide closely related to nylon is prepared by prolonged heating of caprolactum at 530-540 K. The fibre is practically identical to nylon in properties.

Perkin Reaction: A condensation between aromatic aldehydes and the sodium salts of fatty acids or their aromatic derivatives. The reaction between benzaldehyde and sodium ethanoate in the presence of ethanoic anhydride leads to sodium cinnamate.

$$C_6H_5CHO + H_3C.COONa \rightarrow C_6H_5CH{:}CHCOONa$$

In general, condensation takes place at the a-carbon atom, leading to simple cinnamic acids or to their a substituted derivatives.

Permalloy: Class of iron-nickel alloys with high magnetic permeability.

Permanent Gas: A gas that cannot be liquefied by pressure alone; a gas above its critical temperature.

Permanent Hardness: A kind of water hardness created by the presence of dissolved calcium and magnesium chloride or sulphate.

Permanganic Acid, $HMnO_4$: A hypothetical acid known only in solution or in the form of its salts, the permanganate.

Permutit: The double silicates of sodium or potassium with aluminium. The permutits are represented by the chemical formulae

Sodium permutit $Na_2O.\ Al_2O_3.4SiO_2 2H_2O$

Potassium permutit $K_2O.\ Al_2O_3.4SiO_2 2H_2O$.

It is largely used for removing hardness from water.

Pernicious Anaemia: Lack of adequate number of red blood corpuscles (RBC) in blood., resulting from the inability of bone marrow to produce normal RBC due to deficiency of cyanocobalamine (vitamin B_{12}).

Pernitric Acid (HNO_3): Obtained by dissolving nitrogen pentaoxide in hydrogen peroxide at low temperatures.

$$N_2O_5 + H_2O_2 \rightarrow HNO_3 + KNO_4.$$

The potassium salt of pernitric acid is prepared by cooling a concentrated solution of potassium nitrate is a freezing mixture and adding to it sodium peroxide gradually and alcohol.

$$KNO_3 + Na_2O_2 + H_2O \rightarrow 2NaOH + KNO_4.$$

Attempts to isolate have not been successful.

Peroxosulphuric (V1) Acid: A white crystalline solid formed by reacting hydrogen peroxide with sulphuric acid. Also called Caro's acid (H_2SO_5), it is a powerful oxidizing agent.

Perspex: A widely used acrylic resin, polymethyl acrylate.

Pesticides: Substances that kill pests; they include insecticides.

Petrochemicals: Chemical substances derived from petroleum (or natural gas).

Petrol (Gasoline): A complex mixture consisting of hydrocarbons, such as, hexane, heptane and octane.

Petroleum: A complex mixture of hydrocarbons formed over a period of millions of years by decomposition of plant matter.

Petroleum Ether: A mixture of the lower hydrocarbons of the paraffin series consisting mainly of pentane and hexane.

pH: The logarithm of the reciprocal of the hydronium ion concentration.

pH Meter: An instrument which is employed to determine the pH of solution.

Phase: A distinct portion of matter that is uniform throughout in composition and in intrinsic properties.

Phase Diagram: A graphical representation of the state in which a substance will occur at a given pressure and temperature.

Phase-Equilibrium: A part of the system which is physically homogeneous and is separated from other similar parts of the system by surfaces of separation. It is different from them in state or composition or both and is in dynamic equilibrium with them through exchange with one or more chemical species.

Phase Rule: J. Willard Gibbs provided a general and useful relationship between the number of phase P, the number of components, C and the number of degree of freedom, F which must be specified in order to characterise a system. Mathematically, it is written as

$$P+F=C+2 \text{ or } F=C-P+2$$

Phenanthrene ($C_{14}H_{10}$): A white solid (m.p. 372 K) isolated from anthracene oil, a fraction of Coal-Tar collected between 540 and 630K. It is insoluble in water but soluble in alcohol, ether and benzene.

Phenazine, $C_3H_4N_2C_6H_4$: A yellow crystalline substance, m.p. 171°C, used in the manufacture of dyes.

Phenetole, $C_6H_5OC_2H_5$: Phenyl ethyl ether. A volatile aromatic liquid, b.p. 172°C.

Phenobarbitone: Luminal Phenylcthylbarbituric acid $C_6H_5.C_2H_5C:(NHCO)_2:CO$, A white crystalline powder, m.p. 174°C; used as a sedative and hypnotic drug usually in the form of the soluble sodium salt.

Phenolphthalein: A white crystalline solid insoluble in water but soluble in alcohol. With alkalies it gives a pink colour which disappears on adding an acid. It is used as indicator.

Phenyl Group: Benzene derived C_6H_5—group.

Phenyl Hydrazones: Derivatives of phenyl hydrazine from aldehydes and ketones, with common structure $RCH:NNC_6H_5$ or $RRC: NNHC_6H_5$, where R and R′ are alkyl or aryl groups. These compounds have characteristic melting points which help in qualitative analysis of aldehydes and ketones.

Phenyl Radical (C_6H_5): A radical formed from benzene by the removal of hydrogen atom.

Phenylbutazone (Trade name Butazolidine): A pyrazolone derivative with structural formula as given below :

Phenylethene (Styrene, $C_6H_5CHCH_2$): A colourless liquid obtained in the laboratory by beating cinnamic acid with a small amount of quinol. It is used for making synthetic rubber and plastics.

Phenyl Methanol (Benzyl Alcohol, (C_6CH_2OH): A colourless liquid (b.p. 205°C) obtained by hydrolysing benzyl chloride with aqueous solution hydroxide

$$C_6H_5CH_2Cl + NaOH \rightarrow C_6H_5CH_2OH + NaCl$$

Pheromone (Sex): Chemicals produced or secreted by animals of one sex to induce mating in the opposite sex. For example, propylure secreted by female pink boll worm moth is used to attract male moths in field traps.

Phillips Process: A method for the manufacture of high density polyethylene using a catalyst of chromium (III) oxide on a promoter of silica and alumina.

Phlogiston Theory: A former theory of combination. Flammable compounds were supposed to Contain a substance (phlog) which was released when they burned, leaving ash.

Phonon: The quantum of thermal energy in the lattice vibrations of a crystal. If *f* is the vibrational frequency, the magnitude of the phonon is *hf*, where *h* is Planck's constant.

Phosphate: Salt or ester of phosphoric acid, H_3PO_4; e.g., Sodium phosphate, Na_3PO_4; triethyl phosphate $(C_2H_5O)_3PO$.

Phosphide: A compound of phosphorous with a more electropositive element.

Phosphine (PH_3): A colourless gas obtained by treating phosphide with water (calciphosphide or sodium phosphide or aluminium phosphide). It is highly poisonous in nature. It is used in obtaining smoke screen and as Holme's signal for ships.

Phosphinic Acid (Hyphophosphorus Acid, H_3PO_2): A colourless crystalline solid, obtained when white phosphorus is heated with analkali solution. The alkali salt of hypophosphorus acid is obtained and then it is treated with calculated quantity of dilute sulphuric acid. It is readily soluble in water and is a monobasic acid. Some of its salts, e.g., sodium, potassium, calcium are used as nerve tonics.

Phosphite: Salt or ester of phosphorus acid, H_3PO_3; e.g., aluminium phosphite ($AlPO_3$).

Phosphonic Acid (Phosphorus Acid, Orthophosphorus Acid, H3PO3): A white deliquescent solid obtained by dissolving phosphorus trioxide in cold water.

$$P_4O_6 + 6H_2O \rightarrow 4H_3PO_3$$

It is very soluble in water. It is dibasic acid.

Phosphoric Acid (Orthophosphoric Acid, H_3PO_4): A colourless syrupy liquid, which on further concentration by evaporation in Vacuo deposits transparent prismatic, m.p. 42°C. It is a weak tribasic acid. It is prepared by treating phosphorite rock with dilute sulphuric acid.

$$Ca_3(PO_4)_2 + 3H_2SO_4) \rightarrow 3CaSO_4 + 2H_3PO_4.$$

It is used in the manufacture of phosphates which are employed as fertilizers.

Phosphorus: A nonmetallic solid having two allotropic forms white and red. It is the second element of group V of the periodic table with electronic configuration $[Ne]_3S_23P_3$. White phosphorus is a waxy, soft transparent solid that is about twice as dense as water. It is soluble in carbon disulphide and insoluble in water. It is very poisonous in nature. Red phosphorus is insoluble in water and carbon disulphide. It is used in the preparation of alloys, matches, phosphoric acid and phosphorus acid.

Symbol P; m.p. 15; m.p. 441°C (white P); b.p. 280.1°C; r.d. 2.70.

Phosphorus (III) Bromide or Phosphorus Tribromide (PBr3): A colourless liquid (b.p. 173°C) obtained by adding a solution of bromine in benzene or carbondisulphide to dry phosphorus. It is hydrolysed by water into phosphorus acid and hydrobromic acid.

$$PBr_3 + 3H_2O \rightarrow H_3PO_3 + 3HBr$$

It closely resembles the phosphorus trichloride in its properties.

Phosphorus (V) Bromide or Phosphorus Penta Bromide (PBrs5): A yellow solid (b.p. 106°C) formed by the action of bromine on phosphorus tribromide. It on heating decomposes to give phosphorustribromide and bromine. It is hydrolysed by water. It resembles PCl_5 in its reactions, but it is less vigorous.

Phosphorus (III) Chloride or Phosphorus Trichloride (PCl_3): A mobile liquid (b.p. 74°C) obtained by the action of dry chlorine on yellow phosphorus. It fumes in moist air and reacts with water violently to give phosphorus acid :

$$PCl_3 + 3H_2O \rightarrow H_3PO_3 + 3HCl$$

It reacts with organic compounds containing hydroxyl group (-OH), e.g., (ethyl alcohol) ethanol. $C_2H_5OH + PCl_5 \rightarrow C_2H_5C1 + H_3PO_3$

Phosphorus (V) Chloride or Phosphorus Pentachloride (PC15): A pale yellow solid obtained by the action of excess of chlorine on phosphorus trichloride. It fumes in air and has pungent smell. It sublimes below 100°C. It reacts violently with water With a hissing sound. It reacts with organic compounds.Containing hydroxyl group. It is used in organic reactions to replace hydroxyl group.

Phosphorus (III) Iodide or Phosphorus Triodide (PI_3): A red crystalline solid (m.p. 61°C) obtained by reacting phosphorus (1 part) with iodine (12.3 part) in carbon disulphide solution and cooling the mixture in a freezing mixture. It dissociates when heated. It is hydrolysed by water to give phosphorus acid and hydroiodic acid

$$PI_3 + 3H_2O \rightarrow H_3PO_3 + 3HI$$

Phosphorus (III) Oxide or Phosphorus Trioxide: A white solid with a garlic odour, m.p. 24°C and b.p. 173°C. It readily dissolves in water to form phosphorus acid (H_3PO_3). It is used in the preparation of phosphorus acid and phosphites.

Phosphorus (V) Oxide or (Phosphorus Pentaoxide P2O5): A white solid which forms solution on beating. It forms phosphorus acid (H_3PO_4) with excess of water.

Phosphoryl Chloride ($POCl_3$): Obtained by the hydrolysis of phosphorus pentachloride with a calculated quantity of water.

$$PCl_5 + H_2O \rightarrow POCl_3 + 2HCl.$$

It is a colourless fuming liquid (b.p. 380 K) with a pungent odour. It reacts with finely divided metals at higher temperatures to form the corresponding chlorides.

$$6Ag + 4POCl_3 \rightarrow Ag_3PO_4 + 3AgC1 + 3PCl_3$$

Phot: A unit of illumination in the e.g.s. system equal to an illumination of one lumen per square centimetre.

Photocathode: A cathode that emits electrons when it is illuminated, i.e. as a result of the photoelectric effect.

Photo Chemical Smog: A complex mixture of undesirable substance formed by the action of sunlight on an urban atmosphere polluted with automobile emissions.

Photodissociation: The process which occurs when a molecule on absorbing a quantum of light energy undergoes dissociation to smaller molecules radicals or atoms, e.g., propanone undergoes photodissociation to methyl radicals and carbon monoxide.

Photo-Electric Effect: The emission of electrons from the surface of metals when metal is irradiated by light.

Photo-Electrochemical Cell: A type of voltaic cell in which one of the electrodes is a light-sensitive semiconductor, causing electrons to pass between the electrode and the electrolyte and produce chemical reaction in the cell.

Photo-Electron: An electron ejected from a solid, liquid or gas by photoelectric effect.

Photo Synthesis: The conversion of carbon dioxide and water into carbohydrates in the presence of light by plants. The green colouring matter in plants, chlorophyll acts as a catalyst in the reaction.

Physical Chemistry: The branch of chemistry concerned with physical properties of compounds and how these depend on the chemical bonding.

Physiological Solution: An isotonic solution of salt (NaCI, 0.085%) in distilled water, used for preserving cells.

Phytotonic Chemical: A substance that increases the quality of plants, e.g., increase in root or shoot length, chlorophyll content in leaves, fruit growth etc., i.e., a growth regulator.

Physostigmine (Eserine $C_{15}H_{21}O_2N_3$): A colourless alkaloid, m.p. 105-6°C, used in the treatment of glaucoma.

Piezo Electricity: The generation of an electric charge by certain crystals when subjected to pressure.

Pig Iron: Crude iron produced from a blast furnace. Containing carbon, silicon and other impurities.

Pinch Effect: (1) The constriction of a liquid conductor of electricity (e.g., mercury or molten metal) that occurs when a substantial current is passed through it. (2) The constriction of a plasma due to the magnetic field of a high current within the plasma.

Plane Polarization: A type of polarization of electromagnetic radiation in which the vibrations take place entirely in one plane.

Plasmolysis: Flowing out of plasma from a cell by osmic pressure and causing contraction of cell, i.e., if a cell is placed in a solution of a greater molecular concentration (hypertonic solution) than the contents of the cell, the water in the cell flows out through cell walls causing plasmolysis and contraction of the cell.

Plaster of Paris ($CaSO_4.H_2H_2O$): A gypsum ($CaSO_4.2H_2O$) on heating between 120°C to 130°C gives plaster of paris.

$$2CaSO_4.2H_2O \rightarrow (CaSO_4.1/2H_2O + 3H_2O$$

It forms white solid mass when mixed with water. It is used for making statues and surgical treatment. B and age dipped in wet plaster of pail are used in surgery, in order to maintain limbs in a fixed place. It is also used in making the chalk crayons.

Plasticizer: A nonvolatile liquid added to paints and varnishes to prevent brittleness of the dried film.

Platinum: A silvery white transition element. It is chemically inactive. It is used as a catalyst in contact and Ostwald process, as alloy and used for jewellery. Symbol Pt; a.n. 78; r.d. 21.5, m.p. 1770°C; b.p. 3830°C.

Platinum Black: A black powder formed when platinum is precipitated from solution of its tetrachloride by reducing agents like sodium formate, glucose etc. It is widely used as catalyst.

Platinum (II) Chloride ($PtCl_2$): A brown powder obtained by heating chloroplatinic acid to 580° or by heating spongy platinum in a current of chlorine

$$H_2PtCl_6 \rightarrow PtCl_2 + 2HG + Cl_2$$

It is insoluble in water but soluble in hydrochloric acid.

Platinum (IV) Chlorine (PtC_4): A brown solid obtained by beating chloroplatinic acid to 350° in chlorine.

$$H_2PtCl_6 \rightarrow 2HO + PtCl_4$$

It is hygroscopic and dissolves in water giving strongly acidic solution. It is an oxidizing agent and liberates iodine from potassium iodide. It gives chloroplatinic acid, Ht_2Cl_6 with hydrochloric acid.

Platinum Iridium: An alloy of platinum containing upto 30% iridium. It is used in jewellery, electrical contacts, and hypodermic needles.

Platinum Metals: The group of transition metals ruthenium (Ru), osmium (Os), rhodium (Rh), iridium (Ir), palladium (Pd), and Platinum (R).

Plumbane (Lead hydride, PbH_4): A colourless unstable gas that can be obtained by the action of acids on a mixture of magnesium and lead pellets.

Platinum (II) Oxide or Platinum Monoxide or Platinous Oxide (PtO): A black powder obtained by gently heating the hydrated oxide precipitated by adding dilute alkali to a solution of platinous chloride.

$$PLCl_2 + 2NaOH \rightarrow Pt(OH)_2 + 2NaCl$$

$$\downarrow \text{heat}$$

$$PtO + H_2O.$$

It is soluble in acids but not in alkalis. It readily oxidized by disproportionates on heating.

$$2RtO \rightarrow PtO_2 + Pt$$

Platinum (II) Sulphide or Platinous Sulphide (PtS): Formed as black precipitate when H_2S passed into a solution of chloroplatinate less soluble than HgS.

Platinum (IV) Oxide or Platinum Dioxide or Platinic Oxide (PtO_2): A black powder is obtained by adding excess of caustic alkali to a solution of chloroplatinic acid the salts of platinic acid are formed. The action of acetic acid on these yields the platmic acid $H_2[Pt(OH)_6]$. On heating this to a temperature below 370 K, dioxide is formed.

$$H_2[Pt(OH)_6 \rightarrow 4H_2O + PtO_2.$$

It is soluble both in acids and alkalis.

Platinum (IV) Sulphide or Platinic Sulphide (PtS2): Formed as black precipitate when H_2S is passed into a hot solution of chloroplatinic acid. It is insoluble in nitric acid but soluble in ammonium polysulphides.

Platinum Sesquioxide (Pt_2O_3. XH_2O): Formed as a brown precipitate when sodium carbonate is added to a solution of platinum trichloride. It can not be dehydrated.

Plutonium: A highly toxic radioactive silvery element of the actinoid series of metals. It is a transuranic element found on earth only in minute quantities in uranium ores but readily obtained as 239Pu by neutron bombardment of natural uranium.

Symbol Pi; m.p. 641°C; b.p. 3232°C.

PMR (1HNMR): Nuclear magnetic resonance spectroscopy using proton (^1H) nuclei. In a PMR spectrum the protons having identical electronic or chemical environment resonate at one particular radio frequency and produce absorption signals or peaks at a particular chemical shift upfield (towards) or downfield (away) from the internal standard (tetramethyl silane, TMS); the number of peaks being n + 1, where n is the number of protons on neighbouring carbon atoms (or atoms, other than H), A single peak is called a singlet, a two peak signal is known as a doublet, a three peak signal a triplet, a four peak signal a quartet and any group of more peaks a multiplet. The spacing between two adjacent peaks in a doublet, triplet or quartet is called coupling constant or J value and is measured in cycles pr second (Hertz, Hz or cps). Some chemical shifts (in d ppm) for a few important proton signals are 7.25 ($CHCl_3$); 7.2 (C_6H_6); 4.0 (CH–CI), 3.8 (OCH_3); 3.0 (ArCH); 2.3 ($ArCH_3$); 1.6 (CH_3C=C) 0.9 (CH_3—C); and 0.2 (CH_4).

Pnicogen: A collection term sometimes used (but not recommended) for the elements nitrogen, phosphorus, arsenic, antimony, and bismuth.

Point Defects: The defect caused in crystals by missing or misplaced ions. One kind of defect consists of a cation which has been shifted from its proper place to a place between regular lattice sites.

Polar Bond: A covalent bond in which the bonding electrons are not shared equally between the two atoms. These are, actually, the bonds between atoms of different electronegativities and hence there is always a charge separation as in case of hydrogen chloride molecule, $H^{d+} : Cl^{d-}$.

Polar Compound: A compound which has molecules with a permanent dipole moment, e.g., hydrochloric acid (HCl); water (H_2O) etc. Such compounds are hydrophilic in nature. Distribution of a polar substancesis favoured in a polar solvent more than in a lipophilic or nonpolar solvent.

Polar Covalent Bond: A covalent bond in which more than half of the bond's negative charge is concentrated around one of the two atoms.

Polarimeter: An instrument used to measure the amount by which the plane of polarization of planepolarized light is rotated on passing through an optically active medium.

Polar Molecule: A molecule in which the individual polar bonds are not perfectly symmetrically arranged and are therefore not in balance. Thus the charge separation in the bonds gives rise to an overall charge separation in the molecule for example in water.

Polarography: A highly developed electrochemical method of analysis, particularly suitable for dilute solution of substances which are susceptible to electrolytic reduction at a mercury cathode. A dropping-mercury electrode (DME) is used and the polarogram, the current voltage curve, for the solution is recorded. This polarogram shows a step for each reducible species present in solution. This step is of characteristic potential and of height proportional to the concentration of the component, In favourable cases several components can be detected and determined quantitatively from the polarogram. The method is particularly useful for determining trace amounts of metal and it can be used for investigations of complexes in solution and for bodies in nonaqueous solvents.

Polaron: An excitation in a solid consisting of polar molecules resulting from the interaction between an electron and its strain field.

Polonium: A radioactive metallic element belonging to group VI of the periodic table. It occurs in very minute quantities in uranium ores. Symbol Po; m.p. 254°°C; b.p. 962°C.

Polyamide: A synthetic polymer in which the monomers are linked by the group—NH—CO—, Nylon is an example of a polyamide.

Poly Tetrafluoroethene: Synthetic polymer obtained by addition polymerization of tetrafluoroethene. It can withstand high temperature without decomposition, has a low coefficient of friction and is inert to caustic chemical and strong acids. Therefore it is a valuable material for use in laboratory apparatus, solvent bottles, non stick pans, bearings..

Polyatomic: Molecules, ions or radicals containing more than two or several atoms, e.g., methane, CH_4, ethane, C_2H_6; benzene, C_6H_6; ammonium sulphate, $(NH_4)_2SO_4$ etc.

Polyatomic Ions: A group of atoms joined together as in a molecule but having a net positive or negative charge, e.g., NO_3^-, SO_4^{2-}, PO_4^{3-}.

Polychloroethane (Polyinyl Chloride, PVC): A synthetic polymer made from chloroethane. It is a strong material with a wide variety of uses.

Polyerystalline: A substance composed of many minute interlocking crystals that have solidified together.

Polyester: A synthetic polymer made by reacting alcohols with acids so that the monomers are linked by the group—O—CO—.

Polyethene: A synthetic polymer made from ethene. Produced in two forms—a soft material of low density and a harder, higher density, rigid form.

Polymer: A compound made up of large or very large molecules (macromolecules) formed by condensation of a number of similar molecules (monomers) in a linear fashion linear polymers) or in a net like structure (cross baked polymers). Polymers do not usually have a definite relative molecular mass because of variation in their skeletal chain lengths. They may be natural substances (e.g., cellulose, proteins and starches) or synthetic materials..

Polymerisation: The process that takes place when a single substance (monomer) gives products having the same empirical formula but different molecular weight, each of these being a multiple of that of the monomer.

Polymorphism: The ability of certain chemical substances to exist in more than one physical form. The existence of two forms is called dimorphism.

Polypropylene (Polypropene): A synthetic polymer made from propene. It is similar in properties to polythene but stronger and lighter.

Polyprotic Acid: Acid that can lose more than one proton per molecule, e.g., phosphoric acid (H_3PO_4) is a tripotic acid, sulphuric acid (H_2SO_4) is a dripotic acid.

Polysaccharide: Higher molecular polymers of the sugars. They contain many repeated units in their molecular structures. They can be brokendown to smaller polysaccharides, disaccharides and monosaccharides by hydrolysis.

Polystyrene: A synthetic polymer made from styrene. Expanded polystyrene is a rigid form used in packing and insulation.

Polyurethane: A synthetic polymer containing the group —NH—CO— linking the monomers. Polyurethanes are made by condensation of isoeyanates with alcohols.

Polyvinyl Chloride (PVC): A colourless thermoplastic material, produced by the polymerization of vinyl chloride (CH_2CHCI), with good resistance to water, acid, alkalis and alcohols.

Portland Cement: A mixture of tricalcium silicates ($3CaO.SiO_2$) dicalcium silicate ($2CaO.SiO_2$) tricalcium aluminate ($3CaO.\ AI_2O_3$). It is the most important building material. The above mixture on mixing and heat setting to a hard mass resembling the famous rocks of Portland of England and hence, the mixture is named Portland cement.

Positron: Positive electron An elementary particle with the same mass as the electron and an electric charge of equal magnitude but opposite sign. Positrons are produced during several decay process and during pair production.

Potassium: A silver white alkali metal obtained by the electrolysis of potassium hydroxide or potassium fluoride or potassium cyanide and calcium carbide. It exists in nature as carnallite, KCl, $MgCl_2.\ 6H_2O$; Orthoclase, $K_2O.\ AI_2O_3.6SiO_2$ and mica, $K[Al_3Si_3O_{10}(OH)_2]$. It closely resembles sodium in chemical properties and comparatively more reactive than it. It is used in making alloys and in photoelectric cells.

Symbol K; a.n. 19; e.d. 0.86; m.p. 63.7°C b.p. 756°C.

Potassium Bicarbonate (Potassium Hydrogen Carbonate, $KHCO_3$): A white solid obtained by passing carbon dioxide through a cold solution of patassium carbonate. It is soluble in water and its solution is alkaline.

Potassium Bromide (KBr): A white crystalline solid obtained by saturating a hot solution of caustic potash with bromine.

$$6KOH + 2Br_2 \rightarrow 5KBr + KBrO_3 + 3H_2O$$

The solution is evaporated to dryness and further heated with charcoal to reduce the bromate to bromide

$$KBrO_3 + 3C \rightarrow KBr + 3CO$$

It is fairly soluble in water and solution can dissolve bromine. It is mainly used in photography in preparing photographic plates and also used in medicine as sedative.

Potassium Carbonate (Pearl Ash, K_2CO_3): A white hygroscopic solid, m.p. 900°C. It is highly soluble in water giving an alkaline solution due to hydrolysis.

Potassium Chlorate ($KClO_3$): A fine white flake sparingly soluble in water. It is obtained by the electrolysis of sodium chloride solution at 40°C. The solution thus obtained is treated with potassium chloride

$$6NaOH+3Cl_2 \rightarrow 5NaClO_3+3H_2O$$

$$NaClO_3+KCl \rightarrow KClO_3+NaCl$$

It is used as powerful oxidising agent in making matches and fire works.

Potassium Chloride (KCl): A white crystalline solid, fairly soluble in water. It is found in nature as sylvine, KCI and carnallite, $KCl.MgCl_2.6H_2O$. It is used as fertilizer. It is as used in the preparation of various compounds like KOH, K_2CO_3 and KNO_3.

Potassium Cyanide (KCN): A white crystalline solid, m.p. 635°C. It is highly hygroscopic and very soluble in water. It is a deadly poison. It is used in electroplating and extraction of gold and silver.

Potassium Ferrocyanide [$K_3Fe(CN)_6$]: A dark red crystal, readily soluble in water. This salt is prepared by the oxidation of potassium ferrocyanide by means of chlorine

$$2K_4[17e(CN)_6 + Cl_2 \rightarrow 2K3[17e(CN)_6] + 2M$$

It is used to identify Fe_2, ion and as external indicator in the titration.

Potassium Hydroxide (Caustic Potash, KOH): A white highly deliquescent solid obtained by the manufacture of soft toilet soaps and as an electrolyte in the nickeliron storage battery. An alcoholic KOH is used in many organic reactions.

Potassium Hydrogen Tartrate (Cream of Tarlar, $C_4O_6H_5K$): A white crystalline powder obtained from argol (tartar), used in baking powder.

Potassium Iodide (K1): A white crystal, m.p. 680°C. It is soluble in alcohol and water. It is prepared by adding iodine to a hot concentrated solution of caustic potash forming the iodine and iodate. The solution is evaporated to dryness, the residue is heated with charcoal to reduce the iodate to the iodide. It gives scarlet precipitate of HgI_2 with mercuric chloride which is soluble in excess of KI.

$$2KI + HgCl_2 \rightarrow HgI_2 + 2KCl$$

$$2KI + HgI_2 \rightarrow K_2[HgI_4]$$

An alkaline solution of $K_2[HgI_4]$ is known as Nessler's reagent. It is mainly used in photography.

Potassium Manganate (vii) (Potassium Permangnate, $KMnO_4$): A dark purple, anhydrous needle like crystal with a metallic lustre. It is obtained by the electrolytic oxidation of the manganate. It is used as an oxidising agent, disinfectant and in organic chemistry as Bayer's reagent.

Potassium Nitrate (Saltpetre, Nitre, KNO_3): A white needle like crystal. It is very soluble in water. It is not hygroscopic. It is obtained when a sodium nitrate is added to hot saturated solution of potassium chloride. It is used in the production of nitric acid in gunpowder and fireworks.

Potassium Sodium Tartrate (Rochelle salt, COOK.(CHOH)2COONa,$4H_2O$): A white crystalline soluble salt, m.p. 70-80°C, used in the preparation of baking powder, seidlitz powder, etc.

Potassium sulphate (K_2SO_4): A colourless hygroscopic substance, m.p. 173.2°C, used in the manufacture of dyes and drugs.

Potential Energy: The energy stored in an object by virtue of its position or composition.

Potentiometric Titration: A titration in which an electrode is used in the reaction mixture. The end point can be found by monitoring the potential of this during titration.

Prasedodymium: The soft ductile malleable silvery element of lanthanoid series of metals. It is used in the preparation of alloys, as a catalyst and in the glass industry.

Symbol Pr; a.n. 59; r.d. 6.77; m.p. 931°C; b.p. 3212°C.

Prechill: A technique to break dormancy in seeds. Seeds are placed in a moist substratum so that they can take up (imbibe) water and then held at a low temperature 5-10°C) for a period usually 24 days, before transferring to normal germination temperature. For some tree seeds, long term preehilling is required.

Precipitate: An insoluble substance formed in a solution as the result of a chemical reaction.

Precipitation: The formation of a precipitate A common type of precipitation much used in chemical analysis and preparations, occurs by double decomposition when two solutions are mixed if each of the solutions contains one radical of an insoluble compound.

Pressure: The force per unit area of surface that a gas or liquid exerts. It may be expressed in g/cm^2 or simply as height of mercury in centimetres, millimeters or inches. The Torr, a unit of pressure, equal to 1/760th of the normal atmosphere or to 1 mm of mercury.

Printing Inks: Normal mixtures of pigment (e.g., carbon black) dispersed in an oil of a suitable viscosity.

Probit Analysis: A statistical procedure of transforming percentage mortality data into another variable called probit and then finding the regression function of probit with log dosage in order to estimate LD_{90} or LD_{90} etc. (of any toxic material).

Producer Gas: A mixture of carbon monoxide and nitrogen prepared by burning coke. It is used as a fuel.

Progesterone, ($C_{21}H_{30}O_2$): A white crystalline steroid hormone, m.p. 128.5°C, responsible for preparing the reproductive organs of mammals for pregnancy and for protecting the embryo.

Proguanil: Having trade name Paludrine It is less toxic and does not colour the skin. It is a suppressive drug of choice for both falciparum as well as vivax malarial.

Proinsecticide: An insecticide precursor that gets converted into an insecticide in *vivo*. N-sulphonylated, N-sulphenylated or N-phosphinothiolated derivatives of an insecticide carbamate can act as proinsecticide. For example, carbosulfan is a proinsecticide which produces carbofuran in vivo and has spectrum of activity similar to carbofuran. Such compounds have advantage of being non hazardous during handling due to low mammalian toxicity.

Promethium: A radioactive element of the lanthanoid series of metals. It is not found naturally but can be produced artificially by the fission of uranium.

Symbol Pm; m.p. 1080°C; b.p. 2760°C.

Prompt Critical: Capable of sustaining a nuclear fission chain reaction on the prompt neutrons alone, without contribution from delayed neutrons.

Prompt Neutrons: Neutrons resulting from nuclear fission (either during the fission process or from freshly formed fission fragments) that are emitted without measurable delay, i.e., in less than a millionth of a second.

Proof Spirit: Ethanol containing 49.28% alcohol by weight, or 5,7.10% by volume, and having a relative density of 0.0919 76 at 60°F. Formerly defined as the weakest solution of alcohol that would fire gunpowder when brought into contact with it and ignited.

Propane (C3H8): The third hydrocarbon of the alkane series. An inflammable gas, b.p. –42.17°C. Used as a fuel.

Propane-1, 2, 3-Triol (Glycerol, $CH_2OHCH.OH.CH_2OH$): A colourless, odourless, syrupy liquid obtained by the hydrolysis of fats and oils with alkalies (NaOH). It is sweet in taste. It is miscible with water but is almost insoluble in ether. It is very hygroscopic in nature. It is used in the preparation of formic acid allyl alcohol, nitroglycerine, and in the preparation of soap and cosmetics.

Propanoic Acid (Propionic Acid, C_2H_5COOH): A colourless oily liquid (m.p. –22°C; b.p. 141°C) obtained industrially by the oxidation of n-propanol.

$$CH_3CH_2CH_2OH \xrightarrow{2(O)} CH_3CH_2COOH$$

It has acrid odour. It is miscible with water, ethanol, and ether in all proportions.

Propanol (Propyl Alcohol, C_3H_7OH): Two isomeric propyl alcohols are known. Propane-1 ol (n-propyl Alcohol) is colourless liquid, miscible with water, ethanol and ether. It is used for toilet preparation, such as, lotion. Propane-2 ol (iso propyl alcohol) is colourless liquid and used to prepare ester, acetone and used as solvent and fuel.

Propanone (Acetone, CH_3COCH_3): A colourless, pungent, smelling liquid b.p. –47°C, miscible with water, ethanol, and ether in all proportion. It is used as solvent, for acetylene, cellulose acetate and nitrate.

Propellant: The explosive substance used to fill cartridges, shell cases of weapons and solid fuel rockets. The term is also used to include the fuel and oxidant of rockets when these are separates.

Propene (Propylene, $CH_2.CH.CH_3$): The second member of the alkene series of hydrocarbons. A colourless gas, b.p. –47°C.

Propogation Step: A step in a chain reaction in which a free radical reacts with a reactant molecule to give a product molecule and another free radical.

Protactinium: A toxic radioactive element of the actinoid series of metals. It occurs in minute quantities in uranium ores. Symbol Pa; m.p. 1565°C.

Proteins: High molecular weight (18000-50,000,00) macromolecules formed by peptide linkages, joining thousands of aminoacids. The sequence ammoacids gives individual proteins their specific properties. The particular sequence of the aminoacids in proteins which are synthesised in cytoplasms of cells, is determined by the sequence of the nucleotides in the nucleic acids of the chromosomes, three nucleotides coding for each aminoacid. Proteins form important constituents of all living cells, enzymes and hair. These are of two types, viz,, simple proteins which give only amino acids on hydrolysis; and conjugated proteins which, contain a prosthetic group along with aminoacids in their molecular structure.

Protocatechaldehyde or 3-4-Dihydroxy Bepzaidehyde: A crystalline solid (m.p. 430 K) obtained by heating Vanillin with hydrochloric acid. It is soluble in water. It gives green colonation with $FeC1_3$ and reduces ammonical silver nitrate.

Protolysis (Propylene, $CH_2.CH.CH_3$): The second member of the alkeneseries of hydrocarbons. A colourless gas, b.P. -47°C.

Proton A nucleon of positive charge (+ 1.602192 coulomb) and rest mass equal to 1.672614 ´ 1027 kilogram. It is found in nuclei of all atoms of all isotopes.

Protonic solvents: Liquid molecules (solvents) containing hydrogen that is attached to oxygen or nitrogen and hence are acidic enough to form hydrogen bonding e.g. water, methanol, ethanol, acetic acid etc.

Pseudohalogens: A small group of inorganic compounds with symmetrical molecules that resemble the halogens in some reactions arid compounds. For e.g., cyanogen $(CN)_2$ has compounds analogous to halogen compounds.

Psychedelic Drugs: The drugs produce visual and auditory hallucinations. The person after taking the drugs sees objects and bear sounds which do not exist. They produce euphoria a sense of well being, overjoy, selfconfidence etc. common drugs are lysergic acid diethylamide (L SD-25) and mescaline.

Psychrometer: An instrument for measuring humidity of the atmosphere by calculating the relative difference of temperature in two identical thermometers, the bulb of one of them being kept wet with a thin film of water.

Pyknometer: An apparatus for determining the density and coefficient of expansion of a liquid. It consists of a glass vessel graduated to hold a definite volume of liquid at a given temperature. By weighing it full of liquid at different temperatures, the variations in density, and therefore the apparent expansion may be found.

Pyramid: A solid figure having a polygon for one of its faces (the base), the other face being triangles with a common vertex.

Pyrene, $(C_{16}H_{10})$: (1) A yellow crystalline polycyclin hydrocarbon, m.p. 149°C, found in coaltar. (2) Trade name for a fire extinguisher consisting of carbon tetrachloride, CC_{14}.

Pyrethrins: The active substance of pyrethrum (pyrethrins); e.g. pyrethrin 1, pyrethrin 11, jasmolin 1, jasmolin 11, cinerin 1 and cinerin 11. These are neurotoxins characterized by their quick knockdown effect; but because of unsatisfactory stability these are not used for agricultural purpose.

Pyrethroids, Synthetic: A group of synthetic insecticides similar in structure and activity to natural pyrethrins, e.g., allethrin, permethrin, decamethrin, fenvalerate, fluvalinate etc. They are broad spectrum insecticides required in very small amounts (less than half kilogram per hectare) for effective pest control.

Pyrex: A type of glass that is resistant to heat and chemical attack. It is widely used in laboratory glassware.

Pyridine (C_5H_5H): A colourless liquid (b.p. 11.5°C) with a disagreeable odour. It occurs in the light oil fraction of coal tar and in bone oil. It is used as catalyst, as halogen carrier and as solvent.

Pyro: (1) Prefix denoting strong heat or fire e.g., pyrolysis means decomposition by strong heat. (2) Prefix denoting that the substance is a product obtained by heating another substance e.g., pyroboric acid obtained by heating boric acid.

Pyrogallol: Pyrogallic acid, 1, 2, 3-trihydroxybenzene, $C_6H_3(OH)_3$. Awhite crystalline soluble solid, m.p. 132°C, that is a powerful reducing agent; alkaline solution rapidly absorbs oxygen. Used in photographic developing and in gas analysis for the estimation of oxygen.

Pyrolidine or Tetrahydro-pyrrole (C_4H_9N): Almost colourless liquid (b.P.8889' Q; which fumes in air. It occurs tobacco leaves.

Prophoric: (1) Substance that catches fire in air. (2) A metal or alloy that sparks or produces parks when struck, Lighter flints are made of pyrophoric alloys.

Pyroligneous Acid: A watery liquid obtained by the destructive distillation of wood. It contains acetic acid, CH_3COOH, Methanol, CH_3OH, acetone $(CH_3)_2CO$, and small amounts of other organic compounds.

Pyrolysis: The decomposition of chemical compounds by subjecting them to very high temperature.

Pyrometer: An instrument which is used in chemical industry to determine high temperature.

Pyrophosphate: Salt or ester of pyrophosphoric acid, $H_4P_2O_7$, e.g., tetraethyl pyrophosphate, TEPP.

Pyrophosphoric acid or Dipolyphosphoric acid ($H_4P_2O_7$): Obtained by heating orthophosphoric acid to 523-533 K.

$$2H_3PO_4 \rightarrow H_4P_2O_7 + H_{2)}$$

It is a colourless solid (m.p. 334 K). Pyrophosphoric acid responds to ammonium molybdate test which is a characteristic for orthophosphates.

Pyrrole or Azacycly penta-2, 4-diene (C_4H_5N): A colourless liquid (b.p. 404 K) obtained by passing a mixture of furan, ammonia and steam overheated aluminia.

Q

Quadruple Point: The unique condition of temperature and pressure which four phases of a two component system exist in equilibrium.

Qualitative Analysis: Analysis done with the aim of identifying the components of a sample.

Quantitative Analysis: The analysis done to find the amount of components of a sample. The common methods used are volumetric and gravimetric analysis.

Quanta: The unit of radiation that an atom absorbs.

Quantized: A quantity is said to he quantized if, in accordance with quantum mechanics, it can only have certain discrete values. Such a quantity cannot vary continuously, differences in value being separated by jumps.

Quantum: A definite amount of energy released or absorbed in a process. Energy often behaves as if it were 'quantized' in this way. The quantum of electromagnetic radiation is the photon.

Quantum Electro Dynamics: The use of quantum mechanics to describe how particles and electromagnetic radiation interact.

Quantum Numbers: The energy levels of an electron may be described by four numbers. The first quantum number n which corresponds to the energy level in which the electron is found. It may have values of 1, 2, 3... and up. The second quantum number l which tells the shape of electron cloud. It will have the values of 0, 1, 2... upto n 1. The third quantum number, m which tells the orientation of electrons. It will have the valules $l, l - 1, l - 2, ... 0, (l - 1), -l$. The fourth quantum number S which tells the spin of electrons. It will have the values + 1/2 and –1/2.

Quantum States: States of an atom, electron, particle etc., specified by unique set of quantum numbers.

Quantum Theory: Max Planck put forward a theory in which he said that the absorption or emission of energy takes place in discrete units of quanta and not in continuous stream.

Quartet: A group of four closely spaced lines in a spectrum (*n.m.r.* etc.)

Quaternary Compounds: A white crystalline compound soluble in water. They can be prepared by the action of ammonia with excess of an alkyl halide or by direct action between a tertiary amine and alkyl halide.

Quenching: The process of terminating the discharge in a Geiger counter by preventing regignition.

Quenching of Steel: Rapid cooling by immersion into water or oil, to harden the steel.

Quinhydrone: $[(C_6H_4OH)_2 . C_6H_4O_2]$ An addition compound of hydroquinone and quinone. A green crystalline substance, m.p. 171°C, used in photography and as an antioxidant; the quinhydrone electrode is used in pH measurement.

Quinidine ($C_{20}H_{24}N_2O_2$): A colourless alkaloid, isomeric with quinine, m.p. 174-5°C, used in medicine.

Quinine ($C_{20}H_{24}N_2O_2$): A colourless bittertasting crystalline alkaloid that occurs in Cinchona bark, m.p. 57°C. It is used in the treatment of malaria.

R

Racemic Modification: An equimolecular mixture of enantiomorphs which has no effect on planepolarized light since it contains an equal number of dextrorotatory and laevorotatory forms.

Racemisation: The conversion of an optical isomer into an equal mixture of isomers which is not optically active.

Rad: A measure of energy absorbed from radiation by tissue or other biological material. 1 rad = transfer of 1×10^{-2} J of energy per kilogram of material.

Radian (Symbol rad): The SI unit of plane angle; 2p radian is one complete revolution (360°).

Radiation: In general the emission of energy from a source, either as waves (light, sound, etc.) or as moving particles (beta rays or alpha rays).

Radical (Radicle): A group of atoms, present in a series of compounds, that maintains its identity through chemical changes affecting the rest of the molecule, but that is usually incapable of independent existence, e.g., the ammonium radical NH_4^- ethyl group, C_2H_5.

Radioactive: Describing an element or nuclide that exhibits natural radioactivity.

Radioactive Dating: Any method of measuring the age of materials that depends on radio activity.

Radioactive Decay Series: The series of isotopes into which a radioactive nucleus is successively transformed.

Radioactivity: The spontaneous disintegration of an unstable atomic nucleus which accompanies emission of alpha, beta or gamma radiations.

Radiochemistry: The study and application of chemical techniques to purification of radioactive materials and the formation of compounds containing radioactive elements.

Radioisotope: A radioactive isotope of an element.

Radiolysis: A chemical reaction caused by high energy radiation like X-rays, gamma rays etc.

Radiowaves: A form of electromagnetic radiator with wavelengths greater than a few millimetres.

Radium: A white radioactive metallic element of the IA group. It gets oxidised with air violently. It reacts with water to liberate hydrogen which readily catches fire. It is used to produce alloys, radiotubes, and photoelectric cells.

Symbol Ra; a.n. 88; m.p. 700°C.

Radium Bromide ($RaBr_2.2H_2O$): Prepared by dissolving radium carbonate in hydrobromide acid and crystallizing out from the solution. On exposure to air, it loses bromine and absorbs moisture and carbondioxide to form radium carbonate.

Radium Carbonate ($RaCO_3$): Obtained by adding ammonium carbonate to a radium salt solution or by fusing radium sulphate and sodiumcarbonate. It is insoluble in water.

Radium Chloride ($RaCl_2.2H_2O$): Prepared by dissolving radium carbonate in hydrochloric acid and crystallizing out from the solution when crystals of $RaCl_2.2H_2O$ isomorphous with $BaCl_2.2H_2O$ are obtained. It loses water at 470 K and melts on further heating to a liquid which solidifies to a glassy mass shining with bluish violet light.

Radium Hydroxide ($Ra(OH)_2$: Obtained when radium reacts with water.

Radium Nitrate [$Ra(NO_3)_2$]: Prepared by dissolving radium carbonate in nitric acid and crystallizing from the solution.

Radium Sulphate ($RaSO_4$): Precipitated as white substance by adding a soluble sulphate to a radium salt solution. It is about hundred times less soluble than barium sulphate.

Radon: A colourless monoatomic radioactive element of zero group. It is obtained by dissolving salt in water and allowing the solution to stand for about four weeks. It is used in radiotherapy.

Raffinose or Melitose ($C_{13}H_{32}O_{16}.5H_2O$): A colourless crystalline trisaccharide (m.p. 60°C). It is found in beet sugar but does not have sweet flavour.

Raman Effect: When light of frequency V_0 is scattered by molecules of a substance which have a vibrational frequency of V_1, the scattered light when analysed spectroscopically has lines of frequency V_1 where

$$V = V_0 \pm V_1$$

This spectrum is named as Raman spectrum.

Raney Nickel: A spongy type of nickel, used as a catalyst especially in the hydrogenation of fats and oils. It is made by dissolving the aluminium in a nickel-aluminium alloy with sodium hydroxide.

Raoult's Law: The vapour of any component of a solution at any given temperature is equal to the product of the mole fraction of that component m the solution and its vapour pressure in the pure liquid salt at the same temperature. Mathematically can be written:

$$P_A = X_A P^\circ_A \text{ and } P_B = P^\circ_B$$

where P_A and P_B are the partial pressure of the vapours, A and B, X_A and XB are the mole fractions of A and B and $P^\circ{}_A$ and $P^\circ{}_B$ are the vapour pressure of pure A and pure B.

Rasch Parameter: The parameter used to express quantitatively the interelectronic repulsion between various energy levels of an atom.

Rate Determining Step: The slowest step in a multistep reaction. The overall rate of a chemical reaction cannot exceed the rate of the slowest step, e.g., in the reaction between acidified potassium iodide and hydrogen peroxide, the first step in which an iodide ion attacks the molecule of hydrogen peroxide is the rate determining step or limiting step

$H_2O_2 + I^- \rightarrow H_2O + Cl^-$ (slow)

$H^+ + OI^- \rightarrow HOI$ (fast)

$HOI + H^+ + I^- \rightarrow I_2 + H_2O$ (fast)

Rate of Reaction: A measure of the amount of reactants that take part a chemical reaction in a unit time.

Reactant: A substance taking part in a chemical reaction.

Reactive Dyes: Dyes that react chemically with the substances being dyed, to form chemical compounds.

Reactor: The largescale chemical processes are done in various environment generally in tanks, tubes, towers or fluidized beds. The term reactor is also used for nuclear reactor.

Reagent: A substance which produces a chemical reaction with a certain chemical and can be used in testing to discover whether that chemicals present, e.g., silver nitrate is a reagent for testing the presence of chloride, bromide and iodide.

Rearrangement: A reaction in which the groups of a compound rearrange themselves to form a different compound.

Recrystallization: The repeated crystallization of a compound to ensure purity of the sample.

Redox: A symbol for oxidation-reduction, e.g., redox potential, oxidation potential.

Redox Indicator: A compound which undergoes a definite colour change during its reversible oxidation or reduction, e.g., methylene blue is colourless when reduced, blue when oxidized.

Redox Reaction: A reaction in which oxidation and reduction go side by side.

Reduced Temperature, Pressure and Volume: Ratios of the temperature, the pressure and the volume to the critical temperature, critical pressure, and critical volume respectively.

Reduction: A reaction which proceeds with the addition of electron; conversion of Fe^{3+} to F^{2+}

$$Fe^{3+} + 1e \rightarrow Fe^{2+}$$

Reducing Agent: A chemical that brings about reduction of another chemical; for example copper can reduce ferric ions to ferrous ions and hence copper is a reducing agent for ferric ions ($2Fe^{3+}$ + Cu ® $2F_2$ + Cu^{2+}).

Reductones: Enediol stabilized by conjugation to a carbonyl or similar group.

Refining: The process of eliminating impurities from a substance or taking out a substance from a mixture.

Reflux: Liquid from partial condensation of vapour which is returned to the top of a fractionating column and allowed to flow down the column countercurrent to the ascending vapour.

Reformatski Reaction: Aldehydes and ketones react with α-bromo-fatty acid ester in the presence of zinc powder to give β-hydroxy esters which may be dehydrated to give α,-β unsaturated esters.

Reforming: The cyclization of straight chain hydrocarbons obtained from petroleum oil by heating under pressure with a catalyst, usually platinum or alumina, e.g., manufacture of methyl benzene from *n*-heptane.

Refractory: A material not damaged by heating to high temperatures. Such materials are made into bricks and used for lining furnaces, etc.

Refrigerants: The substance used for cooling, e.g., liquid nitrogen.

Regelation of Ice: The melting point of ice is lowered by increased pressure; therefore ice near its melting point is melted by sufficient pressure, and solidification or regelation takes place again when the pressure is removed

Regnault's Method: Obsolete method for finding out gas densities by directly weighing of a known volume of gas under known conditions of temperature and pressure.

Relative Atomic Mass: Symbol Ar. The ratio of the average mass per atom of the naturally occurring element to 1/12 of the mass of an atom or nuclide of 12°C. It was formerly called atomic weight.

Relative Density: Symbol d. The ratio of the density of a given substance to the density of some reference substance.

Relative Molecular Mass: Symbol Mr. The ratio of the average mass per molecule of the naturally occurring form of an element of compound to 1/12 of the mass of an atom or nuclide of ^{12}C. This was formerly molecular weight.

Relaxation: The process by which an excited species loses energy and falls to a lower energy level.

Repellent: A substance that can induce the sense of repulsion in insect. Some insects can secrete defensive malodorous highly developed principally to repel predacious enemies. Many monoterpinoid alcohols like citronellol and aldehydes like citral, farnesal are known to repel ants and mosquitoes. DEET is commercial mosquito repellent.

Resolution: The properties of many substances cannot be explained by a single structure using single and double bond. The actual bonding maybe considered in the molecule as a hybrid of two or more coventional form of the molecules called resonance forms, e.g., resonance in benzene molecule.

Resolution (in GC or HPLC): The ratio (R) of the spacing between the peak maxima (Dt) to the width (W) of the neighbouring peaks whose widths are w_1 and w_2 respectively.

Resonance: (1) The oscillation of electrons between atoms within an ion (as in carbon and two oxygen, in carbonate ions and carboxylate ions) or molecule in as a consequence of which no single structure can be to the molecule with certainty and the actual structure is a hybrid of the possible chemical structures. (2) In a nuclear reaction if the energy of an incident particle or photon is equal or near to the value of an appropriate energy level of the compound nucleus, then resonance is said to occur. Thus a resonance neutron is one whose energy corresponds to a particular energy level of a nucleus which will readily absorb it. (3) In a magnetic field of appropriate strength, the atomic nuclei having odd mass number (atomic weight), which tend to he aligned by the externally applied magnetic field, but because of their own angular moments, the nuclei precess about the direction of applied magnetic field. (4) If a small periodic force is applied to a system capable of oscillation, the system is in general set into forced oscillations of small amplitude. As the frequency (f) approaches the natural frequency of the system, the amplitude (f_0) of the oscillation builds up and becomes maximum when it is equal to f_0. The system is then said to be in resonance with the exciting force. Ibis type of resonance occurs in sound waves.

Resonance Ionization Spectroscopy: A method allowing the selection, isolation and observation of as small an amount as a single distinct atom by using a laser to ionize that atom selectively.

Resorcin Yellow: A golden yellow dye obtained by coupling diazatised sulphanilic acid with 1, 3-benzene diol. It is used in dyeing silk.

Retort: (1) A glass vessel consisting of a large bulb with a long neck narrowing some what towards the end. (2) In industrial processes, any vessel in which a chemical reaction or process takes place, especially distillation. (3) In the canning industry, a large autoclave for heating sealed cans by superheated steam under pressure.

Reverberatory Furnace: A furnace with a low roof above the hearth so that the heat from the fire is directed in to the reactants on the hearth. It is used for smelting ores.

Reversible Reaction: A chemical reaction that can move in forward and backward direction.

Rr Value (Retention Factor): The ratio of the distance travelled by the solute from the origin line and the distance travelled by the solvent from the origin line.

Rh: The log of the reciprocal of the partial pressure of hydrogen in equilibrium with a system; thus

$$Rh = -\log p.$$

Rhenium: A silvery white rare transition metal with sp. gr. 114. The metal is fairly resistant to air oxidation at the ordinary temperature. It burns in oxygen at 1000° to form Re_2O_7. It forms a very useful alloy with tungsten which is employed in thermocouples. Pt-Re alloys used for making electrodes.

Rhodium: A soft silver white metal of the transition group. It generally occurs alloyed with platinum. Its alloy with platinum is employed in making scientific apparatus.

Ribosome: Cytoplasmic organelle containing RNA site of protein synthesis in cells.

Ricinoleic Acid ($C_{17}H_{32}$ OHCOOH): A yellow liquid, b.p. 227°C, that occurs in castor oil and is used in the manufacture of soap.

Ring Closure: A reaction in which one part of an open chain in a molecule reacts with another part, so that a ring of atoms is formed.

Rinmann's Green ($ZnCO_2O_4$): A spinel obtained when cobalt nitrate solution is placed on zinc oxide and the mixture heated to redness. The green colour forms a delicate test for Zn.

Roasting: A process of heating the ore strongly in excess of air and is usually applied to sulphide ore.

Rm Value (Martin's Factor): The measure of lipophilicity of a substance expressed as $R_m = \log [1/ R_f) - 1$ when Rf is measured in reverse phase chromatographic experiment under standard conditions.

Rock Salt: A transparent naturally occurring mineral form of sodium chloride.

Rodenticides, Coumarin: A group of chronic rodenticides that consists of derivatives of 4-hydroxycoumarin. 3-(1-Phenyl 2-acetylethyl) 4-hydroxycoumarin (warfarin) and 3-[3-(4-bromo-1-biphenyl-4-yl)-3-

hydroxy-1-phenylpropyl] -4 -hydroxycoumarin (bromadiolone) are two of the most widely used anticoagulant rodenticides belonging to this group.

Rodinal: A photographic developer consisting of an alkaline solution of paraaminophenol, $NH_2C_6H_4OH$, with sodium bisulphite, $NaHSO_3$.

Roentgen: A unit of radiation, used for X-rays and g-rays It is symbolized by R and defined in terms of the ionizing effect on air, i.e., that will produce 2.58×10^{-4} Coulomb of charge in one kilogram of dry air.

Rongalite: A compound of sodium sulphoxylate and formaldehyde, $NaHSO_2$ .HCHO. Used as a reducing agent in dyeing.

Rotenoids: Rotenone and other compounds related in structure to rotenone, found in the roots of tropical plants belonging to the family of Leguminosae, e.g., Derris, Milletia, Tephrosia and others. They are respiratory poisons and inhibit mitochondrial respiration by reducing the activity of $NADH_2$-dehydrogenase.

Rotomer: A carbon-carbon bond (C–C)

Rubber: A natural or synthetic polymeric elastic material.

Rubidium: A soft silvery highly reactive element of the alkali metal group.

Symbol: Rb; m.p. 38.89°C; b.p. 688°C; r.d. 1.53 (solid), 1.47 (liquid); p.n., 37, r.a.m. 85.47.

Rubidium Strontium Dating: A method of dating some rocks, used for specimens over 10^9 years old. It is based on the decay of rubidium-87 (Half-life 5×10^{10} years) to yield strontium 87. An estimate of the sample's age is given by the ratio of the two isotopes.

Ruby: A red form of corundum, Al_2O_3 that owes its colour to traces of Chromium. Used in lasers and as a gem stone.

Rusting: The corrosion of iron in air to form an oxide ($Fe_2O.H_2O$). Rusting is caused by the presence of both oxygen and moisture in air.

Rusting of Iron: The iron gets corroded when exposed to humid air due to the formation of reddish brown scale. The rust may be different in colour from bright yellow, brown, to almost black. It consists of mainly hydrated ferric oxide ($Fe_2O_3.3H_2O$) with little ferrous oxide and basic ferrous and ferric carbonate.

Ruthenium: A transition metal that occurs naturally with platinum. Symbol: Ru; n.p. 23.10°C; b.p. 390°C; r.d. 12.4; p.n. 44; r.a.m. 101.07.

Rutile: A crystalline form of natural titanium dioxide, TiO_2.

Ryanodine: A naturally occurring alkaloid that blocks release of calcium ions, Ca^{2+} from cells and tissue.

S

s-Block Elements: The elements of first two groups in the periodic table, i.e., IA (H, Li, Na, K, Ra, Cs, Fr) and IIA (Be, Mg, Ca, Sr, Ba, Ra); so called because the outermost shell in the electronic configuration of these atoms is ns^1 of ns^2.

Saccharimeter: An apparatus for determining the concentration of a sugar solution by measuring the angle of rotation of the plane of vibration of polarized light passing through a tube containing the solution.

Saccharin or O-Sulphobenzoic Imide: A crystalline solid about 550 times as sweet as sugar. It is nearly insoluble in water but its sodium salt is very soluble. Pure saccharin is obtained in greater yield by oxidising O-toluene sulphonamide with $K_2Cr_2O_7$ and H_2SO_4. It is used as sweetening agent in cheap drinks and by diabetic patients and fat persons.

Safrole($CH_2OHC_6H_4O–C_6\ H_5O_5$): A colourless soluble glucoside, m.p.200°C, used as an antipyretic and analgesic.

Salicylaidehyde or O-Hydroxy Benzaldehyde: A colourless pleasant smelling oily liquid, soluble in water and alkali to give yellow solution. It is obtained by reflexing an alkaline solution of phenol with chloroform at 360 K.

Saline-Alkali Soil: Soil containing excess soluble salts and exchangeable sodium which interferes in the normal growth of plant; it possesses p^H more than 8.5; electrical conductivity of saturation extract more than 4 millimhos per centimetre and exchangeable sodium percentage more than 15.

Salinity: A measure of the salt content of sea water, or brine. It is equal to the weight in grams of dissolved salt present in 1 kg of sea water or brine.

Salinometer: A type of hydrometer used for determining the concentration of salt solutions by measuring their density.

Salol or Phenyl Salicylate (O-HO $C_6H_5COMC_6H_5$): A white solid obtained by heating salicyclic acid with phenol in the presence of phosphoryl chloride.

Salt Bridge: A U-shape glass filled with a concentrated solution of salt which conducts the current between the half cell but prevents the mixing of the solution of the half cells.

Salting Out: Precipitating of dissolved substance by addition of another that decreases its solubility, e.g., soap can be salted out by common salt from in water.

Salt of Lemon or Potassium Quadroxalate: A white soluble, poisonous, crystalline salt. It is used for removing inkstains.

Samarium: A silvery element of the lanthanoid series of metals. It occurs in association with other larithanoids. It is used in the metallurgical, glass and nuclear industries.

Sand Mayer Reaction: A reaction in which benzene diazonium chloride is converted into phenyl halide in the presence of cuprous halide and halogen acid, e.g., conversion of benzene diazonium chloride into chlorobenzene.

Sandwich Compound: A complex in which an atom of a transition element, generally chromium is sandwiched between two parallel benzene rings.

Saponification: The reaction in which an ester is hydrolysed with a hydroxide, e.g., hydrolysis of ester with NaOH.

Saponification Number: The number of milligrams of potassium hydroxide needed for the complete saponification of one gram of the fat or oil.

Sapphire: A natural crystalline form of the blue, transparent corundum (alumina, Al_2O_3); the colour being due to the traces of cobalt or other metals.

Satin White: A mixture of $CaSO_4$ and $Al(OH)_3$, $Al(OH_4)_3$ or alum plus $Ca(OH)_2$ It is used as a pigment and surfacing paper

S-Block: The element of IA and IIA having electrons in their outer S-sub-shell.

Scandium: A light weight silvery element belonging to the first transition series. It is used in high intensity lights and in electronic devices.

Scheele's Green: A bright green precipitate, probably, consisting of cupric arsenite, $Cu_3(A_3O_8)_2.2H_2O$. Used as a pigment and insecticide.

Schiff Base: A kind of substance obtained by reacting an aldehyde or ketone with aryl amine.

Schiff Reagent: A solution of rosaniline in water decolorized with sulphurous acid.

Schotten Baumana Reaction: A method for the preperation of N-phenyl benzanide (benzanilide) in which a mixture of phenylamine (aniline) and sodium, hydroxide is stirred while benzoyl chlorlde is added slowly. The N-phenyl-benzamide can be recrystallized from hot ethanol:

$$C_6H_5NH_2 + C_6H_5COCl \rightarrow C_6H_5NlCOC_6H_5 + HCl$$

Seed: A small crystal added to a gas or liquid to assist solidification precipitation from a solution. The seed, usually a crystal of the substance to be formed enables particles to pack into predetermined positions so that a larger crystal can form.

Schrodinger Wave Equation: The wave equation used in wave mechanics to describe the behaviour of a particle in a field of force. It is based on de Broglie's concept that every moving particle is associated with a wave of wavelength *h/mv* (where h is Planck's constant and m and v are the mass and velocity of the particle). In three dimensions the equation has the form :

$$\nabla_2\psi + (8p^2\, m/h^2)\,(E-U)\,\psi=0$$

where $\tilde{N}_2$ is the laplace operator, ψ is the wave function, E is the total energy and U is the potential energy of the particle.

Schweizer's ·Reagent: The dark blue solution formed by dissolving $Cu(OH)_2$ in concentrated ammonia solution. It is used as a solvent for cellulose.

Scientific Method: Observation, explanation and testing of an explanation by additional experiments.

Scientific Notation: Number represented as a decimal number between 1 and 10 multiple by 10 raised to a power.

Scrubbers: These remove impurity from a gas by washing it with a liquid.

Scrum: Any solid substance, especially dirt or waste which floats on the surface of a liquid.

Second Law of Thermodynamics: Whenever a spontaneous event takes place it is accompanied by an increase in the entropy of the universe.

Second Order Reaction: A reaction in which the rate of reaction is proportional to the product of the concentration of two of the reactants or square of the concentration of one of the reactant; i.e.,

rate = K[A] [B] or

rate = $K[A]^2$

Sedimentation: The process of separating an insoluble solid from a liquid in which it is suspended by allowing it to fall to the bottom of the containing vessel, with or without agitation or centrifuging.

Selenic Acid (H_2SeO_4): A strongly corrosive crystalline acid, m.p. 58° with properties resembling those of sulphuric acid.

Selenium: A metalloid which exists in four well known allotropic forms. The first one is amorphous form having bright red colour, the second one is black, glassy mass, the third one is orange red coloured crystals and fourth one is grey coloured crystalline substance. It remains unaffected in air at the ordinary temperature. It is used in glass industry as decolorizer and as catalyst.

Selenium Cell: A type of photoelectric cell consisting of a layer of selenium covered by a thin transparent layer of gold. Light falling on the cell produces a voltage by the photovoltaic effect.

Selivanoff's Test: To fructose add a dilute solution of resorcinol in dilute HCl. A deep reddishbrown precipitate soluble in alcohol is obtained.

Semicarbazone: The derivative of aldehyde or ketone with semicarbazide (H_2N:-NHCONH$_2$); compound of the type RR′C=CONH$_2$ or RCH=NNHCONH$_2$, generally a crystalline solid with sharp melting point, that can be used to characterize the original aldehyde or ketone.

Semio-Chemicals: Substances that modify behaviour of insects by inducing in them the response for sex, feeding, fear or alarm etc. These are further classified into pheromones, secreted by insects for stimulating the insects of the same species and allelochemicals, the substances used for interspecific communications.

Semipermeable Membrane: A membrane, such as, cellophane or parchment that permits some molecules but not all to pass through it.

Senstitive Crops: Crops which are easily injured by pesticide chemicals; even slight drift could cause a lot of damage.

Sequestration: The formation of a complex with an ion in solution, so that the ion does not have its normal activity. Sequestering agents are often chelating agents, such as edta.

Serpek Process: A method for Fixation of atmospheric nitrogen. Aluminium is made to react with nitrogen to give aluminium nitride which is then decomposed by steam to give ammonia.

Sesame Oil: A yellow oil obtained from sesame seeds. m.p. –6°C, r.d. 0.919, used in the manufacture of margarine and cosmetics.

Sesqi: Prefix indicating one and a half times or in a ratio of 2:3; for example a monoterpene is a C_{10} compound while sesqueterpene is C_{15}; if an oxide is MO, sesquioxide is M_2O_3 and dioxide is MO_2

Shell: A group of electrons that share the same principal quantum number. Early work on x-ray emission studies used the terms, K, L, M and these shells $n = 1$ K-shell; $n = 2$ L-shell; $n = 3$ M-shell.

Sherardizing: A method of plating iron or steel with zinc, to form a corrosion resistant coating. The iron or steel is heated in contact with zinc powder to a temperature slightly below the melting point of zinc. At this temperature the two metals amalgamate forming internal layers of zinc-iron alloys and external layer of pure zinc.

Siemen's Process: A method of steel manufacture by addition of scrapmetal, iron oxides and ferromanganese to molten pig iron.

Side Chain: An aliphatic radical or group attached to a straight chain or to a benzene ring or other cyclic group in the molecule of an organic compound.

Side Reaction: A chemical reaction that takes place to a limited extent at the same time as a main reaction. Thus the main product of a reaction may contain small amounts of other compounds.

Siderite: Natural ferrous carbonate $FeCO_3$. An important ore of iron. A meteorite consisting of metals and metallic compounds.

Sigma Pile: An assembly consisting of a neutron source and a moderator, without any fissile material, which is used to study the properties of moderator.

Sigmatropic Reactions: A concerted reaction of the type X—C—(C=C)$_n$ ® (C=C)$_n$—C—X which a group (R) inigrates with its s bond within a p framework of an ene, a diene or a polyene, The migrationis accompanied by a shift of p electrons a shown in 1, 3-sigmatropic shift..

Silanes: Hydrides of silicon with a general formula $Si_n H_{2n+2}$. They are prepared by action of acids on magnesium silicide (Mg_2Si). They are unstable compounds that catch fire on exposure to air, extensively used for reduction in organic synthesis and preparation of siloxanes and silicones.

Silica Garden: This interesting experiment beautiful illustrates the phenomenon of osmosis through silica gels. Silica garden is produced by placing crystals of various coloured salts.

Silica Gel: A bluish white transparent mass having variable composition $SiO_2.x\ H_2O$. It is a porous mass with a large surface area. It is used as a common deodorant, purification of petroleum oil and as absorbent for many gases like ammonia, sulphur dioxide and nitric oxide.

Silicates: The salts of silicic acid. The common silicate sodium silicate Na_2SiO_3.

Silicol Process: The manufacture of hydrogen by the action of sodium hydroxide solution on silicon.

Silicon: Two allotropic forms, amorphous and crystalline. The amorphous form is brown, by growscopic, soft mass and poor conductor of electricity. The crystalline form is grey lustrous crystal, fairly bard and good conductor of electricity. It combines with nonmetal at high temperature. Amorphous silicon dissolves in hydrofluoric acid, H_2SiF_6. It is used in making magnets, transformer cores, anticorrosion steel and as deoxidizer. Silicon bronze is used for telegraph and telephone wires and also used as semiconductor in transistor.

Silicon Carbide: The pure silicon carbide is colourless, but the impure commercial product has a dark brown colour. Its hardness is next to diamond. It is used as abrasive for polishing metallic surface and for cleaning metallic articles prior to enamelling. It is used as a high class refractory for the inner lining of furnaces.

Silicon (IV) Oxide: Two allotropic forms, crystalline and amorphous. It is insoluble in water and in all acids except hydrofluoric acid which attacks it.

$$SiO_2 + HF \rightarrow SiF_4 + 2H_2O$$

Large quantity of silica or sand are used for making glass, cement, synthetic stone and water glass.

Silk: A thread like substance obtained by the silkworm. It consists of mainly of the proteins, sericin and fibroin.

Silver: A white lustrous metal. It is one of the best conductor of heat and electricity. In malleability and ductile, silver comes next to gold. The important minerals are Argentite, Ag_2S; Pyrargyrite, Ag_3AsS_3; Prousite,Ag_3AsS_3; Horn Silver, AgCl.

Silver Amide ($AgNH_2$): Obtained by the interaction of silver nitrate and potassamide

$$AgNO_3 + KNH_2 \rightarrow AgNH_2 + KNO_3$$

Silver Bond: A covalent bond between two elements which involves one pair of electrons only.

Silver (I) Bromide: A pale yellow precipitate is obtained when a soluble bromide is added to a solution of silver nitrate.

$$AgNO_3 + KBr \rightarrow AgBr \downarrow KNO_3$$

It is used very much in photography for making the sensitive emulsions for plates and films.

Silver Carbonate (Ag_2CO_3): A white precipitate obtained by adding sodium carbonate to silver nitrate solution or by the action of carbondioxide on silver oxide. It dissolves in water with carbon dioxide due to the formation of silver bicarbonate. It decomposes on beating into silver carbon dioxide and oxygen.

$$2Ag_2CO_3 \rightarrow 4Ag + 2CO_2 + O_2$$

Silver (I) Chloride: A white curdy precipitate obtained when hydrochloric acid is added to a solution of silver nitrate.

$$AgNO_3 + HCl \rightarrow AgCl + HNO_3$$

It turns violet in the presence of light. It is practically insoluble in water but readily dissolves in aqueous ammonia due to the formation of complex $[Ag(NH_3)_2]^+$.

Silver Cyanide (AgCN): A white curdy precipitate obtained by adding required quantity of potassium cyanide to silver nitrate solution.

$$AgNO_3 + KCN \rightarrow AgCN + KNO_3$$

It is not a stable compound and changes into metallic silver with water or alkali.

Silver (I) Iodide: A pale yellow precipitate prepared by adding of soluble iodide, solution, to a solution of silver nitrate. It is virtually ammonia and is used in photographic industry.

Silver (I) Nitrate (Silver Nitrate, $AgNO_3$): A colourless anhydrous rhombic crystal which is very soluble in water. It is prepared by dissolving silver in dilute nitric acid in stainless steel vessels and evaporating the solution to crystilization. It is used in silvering of glass, laundry marking ink, hair dyes. In the preparation of silver halide emulsion for photographic plates and used as laboratory reagent inqualitative and volumetric analysis.

Silver Permaganate (Ag-MnO_4): Obtained by mixing hot silver nitrate solution with potassium permaganate. It is sparingly soluble in water.

Silver Peroxide (Ag_2O_2): A black powder obtained by the oxidation of silver by ozone. It is not a true peroxide and with acids gives silver salts and oxygen.

$$2Ag_2O_2 + 2H_2SO_4 \rightarrow 2Ag_2SO_4 + 2H_2O + O_2.$$

It is a powerful oxidizing agent.

Silver Sulphate (Ag_2S): A black solid obtained by treating silver or its salts with hydrogen sulphide. It dissolves in alkaline cyanide in the presence of air to form a complex alkali argento cyanide.

$$Ag_2S + 4NaCN \rightarrow 2Na[Ag(CN)_2] + Na_2S$$

This complex is used in electroplating.

Silver Thiocynate (AgCNS): A white precipitate obtained by adding the required quantity of potassium thiocyanate to silver nitrate. It is soluble in excess of potassium thiocyanate.

Siloxanes: Organic compounds of the type R_3Si—O—Si R_3 where R and R′ are alkyl groups. The silicones are polymers of sil oxanes.

Single Bond: A covalent bond between two atoms formed by a single part of electrons by mutual sharing of one electron from each atom. Often reffered to as a sigma bond, (s bond) and represented by a single line as in H–Br

Sizing: The method of filling the pores of paper and giving it some degree of water repellency.

Slag: A glassy material obtained by the combination of flux and gangue in a furnace.

Slip Plane: A plane In a crystal particularly a metal crystal, along which one part of the crystal will move relative to the other part, when a force acts on the crystals. One part of the crystal lattice slips over the other part.

Smelting: The extraction of a metal from its ores by a process involving heat. Generally the process is one of chemical reduction of the oxide of the metal with carbon in a suitable furnace.

Smetic Crystals: Liquid crystals in which the molecules are arranged in layers with their axes parallel and perpendicular to the plane of the layers.

Smog: A dark, thick, dust and sootladen, sulphurous fog that under certain meteorological conditions, pollutes the atmosphere of some industrialcities and the lungs of their inhabitants.

Smoke: A suspension of fine particles of a solid in a gas; smoke from coal consists mainly of fine particles of carbon.

Soda Lime: A grey solid obtained by adding sodium hydroxide solution to calcium oxide, to give a mixture of $Ca(OH)_2$ and NaOH. It is used as drying agent.

Sodamide (NaNH2): A white waxy solid melts at 210°C. It is obtained bypassing dry ammonia gas over heated sodium metal (300°–400°C) in an iron vessel.

$$2Na + 2NH_3 \rightarrow 2NaNH_2 + H_2$$

It is used in explosive industry and in the castner process.

Soda Water: Water containing carbon dioxide, CO_2 under pressure; releasing the pressure lowers the solubility of the gas, and thus causes effervescence.

Sodium: A silvery white metal obtained by the electrolysis of fused sodium chloride, potassium chloride and fluoride mixture, It belongs to IA group. The metal is highly reactive and reacts vigorously with water and halogens, it is used in die preparation of sodium peroxide, sodamide, sodium cyanide and tetraethyl lead. It is used as reducing agent, as deoxidizing agent and acts as a catalyst in the polymerization of isoprene.

Sodium Adsorption Ratio: In any soil the concentration in milliequivalents of sodium ions (Na') divided by half of the square root of the sum of concentrations (in milliequivalents) of calciumn (Ca^{2+}) and magnesium(Mg^{2+}), ions per litre of soil solution.

Sodium Azide (NaN_3): A colourless crystalline substance, used in the manufacture of explosives.

Sodium Benzoate (C_6H_5COONa): A white soluble powder, used as an antiseptic and a food preservative.

SodiumCarbonate (Washing Soda, Na_2CO_3): A white solid melts at 852°C without decomposition. It is obtained by Solvay's process.

An ammonia gas is passed through a 30% solution of brine to get a saturated solution. The impurities of calcium and magnesium get precipitated. Then to the saturated solution of ammonia, carbon dioxide is passed to get the conversion brine to sodium bicarbonate and finally it is heated to obtain sodium carbonate. It is used to glass making, the manufacture of caustic soda by process, in paper making, in watersoftening, in the manufacture of soap, in textile and petroleum refining industries.

Sodium Chlorate ($NaClO_3$): A colourless soluble crystalline substance used in the manufacture of explosives, as a mordant, oxidizing agent, and an antiseptic.

Sodium Chloride (Common Salt, NaCl): A white solid form of cubic crystals which are anhydrous. It melts at about 800°C and volatilises at 1600–800°C. It is used in the manufacture of sodium metal, chlorine, sodium carbonate, and caustic soda. It is an essential constituent of our food and is known as the table salt.

Sodium Cobaltnitrite [Na_2-$(CO(NO_2)_6$]: Obtained by mixing well cooled solution of cobalt nitrate and sodium nitrite and acidifying the mixture with acetic acid. It is an important reagent for the detection and estimation of potassium and ammonium salts with which it gives yellow precipitates.

Sodium Cyanide (NaCN): A colourless solid obtained by heating a mixture of crude calcium, cyanamide and sodium carbonate or sodium chloride with coke

$$CaNCN + Na_2CO_3 + C \rightarrow 2NaCN + CaCO_3$$

$$CaNCN + 2NaCl + C \rightarrow 2NaCN + CaCl_2$$

It is readily dissolved in water. It is used in the extraction of silver and gold. It is also employed in electroplating of gold, silver and copper.

Sodium Cyclamate (C_6H_{11}-$NHSO_3Na$): A white crystalline soluble powder formerly used as a sweetening agent in soft drinks and for diabetics but now hanned from such use owing to possible sideeffects.

Sodium Dihydrogen Phosphate (Sodium Dihydrogen OrthoPhophate, NaH_2PO_4): A white solid formed when a calculated amount of phosphoric acid is added to a solution of the common. It is readily soluble in water. If the salt is sulphate, i.e., Na_2HPO_4. It is readily fused by rapid cooling a glassy mass is obtained which is known as Graham's salt. It is used in baking powder.

Sodium Hydrogen Carbonate: A white solid obtained by passing carbon dioxide through a cold solution of sodium carbonate.

$$Na_2CO_3 + CO_2 + H_2O \rightarrow 2NaHCO_3$$

It is used in the preparation of carbon dioxide, as baking powder and in effervescent drinks. It is also used in medicine to remove acidity of the stomach.

Sodium Hydroxide (Castic Soda, NaOH): A white hydroscopic solid obtained by the electrolysis of sodium chloride using mercury electrode or diaphram cell. It easily dissolves in water with the evolution of water. The dilute solution offers a soapy touch and the solid NaOH are highly corrosive to skin, It is used in making soap dyes, artificial silk and textiles, refining of petroleum and vegetable oils, paper and drugs.

Sodium Hypochlorite (Na-OCl): A white unstable crystalline solid, usually kept in aqueous solution. Used in bleaching paper and textiles and as an oxidizing agent, antiseptic, and fungicide.

Sodium Metaborate Perhydrate ($NaBO_2$, H_2O $3H_2O$): A colourless crystalline salt having some what low solubility. In aqueous solution it behaves like a mixture of borax and hydrogen peroxide. It gives oxygen on heating. It is used in the preparation of disinfectant, in bleaching and soap powder.

Sodium Monoxide (Na_2O): A white solid obtained by heating sodium in a limited supply of moist air at 180°C

$$2Na + O_2 \rightarrow Na_2O_2$$

$$Na_2O_2 + 2Na \rightarrow 2Na_2O$$

Sodium Nitrate (Chile Salt Petre, $NaNO_3$): A white cubic crystals obtained by he neutralization of nitric acid with either sodium carbonate or sodium hydroxide. It is highly soluble in water and is also deliquescent. It exists in nature as chile salt petre and caliche. The former is the purest form of sodium nitrate and latter is an impure form nitrate.

Sodium Nitrate: A pale yellow deliquescent crystal formed by heating sodium nitrate with lead. It is used as reducing agent, in organic synthesis for diazotisation reactions and as qualitative reagent in the laboratory.

Sodium Perborate ($NaBO_3.4H_2O$): A white soluble crystalline substance used in bleaching and as a disinfectant.

Sodium Peroxide (Na_2O_2): A pale yellow solid which turns white on exposure to air on account of the formation of hydroxide and carbonate. It is obtained when sodium is heated in excess of oxygen. It is used for making bleaching bath for silk, wool, hair, horn and rayon. It is also used for preparation of hydrogen peroxide, the purification of the respired air in submarine and diving bells and as powerful oxidising agent.

Sodium Sulphate (Na_2SO_4): A white solid formed by heating sodiumchloride strongly with concentrated sulphuric acid. It is largely used in the manufacture of glass and paper.

Sodium Sulphide (Na_2S): A buff colourless solid and forms several hydrates. It is used for removing hair from hides before tanning and also used in the preparation of dyes.

Sodium Sulphite (Na_2SO_3): A white crystalline soluble powder, used as a food preservative, in bleaching, and in photography.

Sodium Tetraborate: A white soluble crystalline salt, occurring naturally as tincal. On heating it loses water of crystallization and melts to a clear glass like solid. Used as an antiseptic, in fireproofing, as a flux, and in the manufacture of glass and ceramics.

Sodium Thiocyanate (NaSCN): A colourless deliquescent crystalline substance, used in medicine.

Sodium Thiosulphate (Na_2-S_2O_3): A white solid formed by boiling a solution of sodium sulphite with pulverised sulphur until the solution ceases to give alkaline test with litmus paper. It is readily soluble in water and crystallizes as large colourless crystals of the pentahydrate. It is largely used in photography as fixer and for removing stains of silver salts, ad iodine from clothes.

Sodium Vapour Lamp: Luminous discharge obtained by passing an electric current between two electrodes in a tube containing sodium vapour at low pressure, Used in street lighting as the characteristic yellow light is less absorbed by fog and mist than white light.

Soft Soap: A liquid soap made by saponification with potassium hydroxide rather than sodium hydroxide.

Sol: A colloidal solution; term commonly used for colloidal dispersion of inorganic solid in liquids.

Solder: An alloy used in joining metals. The molten solder alloys with the surface to be joined without melting them and solidifies on cooling to form a hard joint.

Solid Solutions: A solid composed of two or more substances mixed together at the molecular level. In such solids atoms, ions, or molecules of one component in the crystal are at lattice positions normally occupied by the other component. Alloys are solid solutions of one metal into another, Crystalline alums are solid solutions of one sulphate in another.

Solubility: The mass of solute per unit volume of solvent of mass of the substance in 100 grams of solvent.

Solubility Product: Many substances are soluble in water to at least small extent. If an insoluble or slightly soluble substance is placed in water an equilibrium is established when the rate of dissolution of ions from the solid equals the rate of precipitation of ions from the saturated solution. So an equilibrium exists between solid silver chloride and saturated solution of silver chloride

$$AgCl(s)\ Ag^{+}(aq) + Cl^{+}(aq)$$

The equilibrium constant is

$$k' = \frac{[Ag+][C]}{AgCl}$$

Since the concentration of a pure solid is constant, [AgCl] may be combined with k' to give

$$Ksp = k'[Agcl] = [Ag^{+l}]\ [Cl]$$

The constant Ksp is called solubility product; the ionic concentrations of the expression are those for a saturated solution at the reference temperature.

Solution: A homogeneous liquid system of two or more species that are intimately dispersed within each other at molecular level. The major component is called the solventn and the minor component is called solute. The dissolution of the solute occurs because of direct intermolecular interaction of the solvent with the ions or molecules of the solute, A part of the energy of thisinteraction appears as change in temperature on dissolution.

Solvation: The attraction of an ion in a solution for molecules of solvent.

Solvent: A liquid in a position to dissolve other materials. The solvent is generally larger component of the solution.

Soxlet Extraction: Apparatus A device for extracting the soluble portion of any substance by a continuous circulation of the boiling solvent through it.

Spatula: A tool shaped like a spoon or which has a flat surface, used for removing small quantities of solids.

Specific Rotation: The rotation of plane polarized light in degrees produced by a 10 centimetre length of a solution containing one gram of a given substance per millilitre of stated solvent. The specific rotation $[a]^{20}_{D}$ is measure of optical activity of a substance in solution measured at using the D-line of sodium.

Spectrochemical Series: Ligands arranged in order of their tendency to cause splitting of the orbital in complexes.

Spectrograph: An instrument for producing a photograph record of a spectrum.

Spectroscope: An apparatus for examining the different wavelengths present in electromagnetic radiations.

Spermaceti: A white, waxy solid consisting mainly of cetyl palmitate, $C_{15}H_{31}CO\text{-}O_{16}H_{33}$. It is obtained from the head of the sperm whale. Used in the manufacture of soaps and cosmetics.

Spin: A property of certain elementary particles where by these act as if these were spinning on an axis, i.e., have an angular momentum. Such particles, because of the electronic charge associated with them also have a magnetic moment and therefore in a magnetic field they align at certain angle to the field direction and precess around this direction; the component of angular momentum around ibis direction being *mh*/2*p* where m is spin quantum number and h is Plarick's constant. For an electron (or proton) m has values + ½ and –½

Spirit: A liquid consisting mainly of ethanol obtained by the distillation of fermented potatoes and fruits.

Sponification Number: The number of milligrams of potassium hydroxide needed for the complete sponification of one gram of the fat or oil.

Spontaneous Reaction: A reaction under a given set of conditions is capable in principle of moving in the forward direction to a considerable extent. It may occur very slowly.

Spray Cone: A theoretical cone of spray formed from the tip of nozzle to the surface of target during spray operation.

Spray Ultra Low Volume (ULV): The spray operation requiring only 1-5 liter per hectare of concentrated pesticide emulsion or solution, sprayed with the help of a ultra low volume sprayer. The operation requires knowledge of equipment wind velocity and its direction and the drift hazards.

Sprayer, Rotary: Any sprayer with a spinning disc that forms small droplets when the spray solution passes onto it. The droplets are carried to the target by the pneumatic pressure or air blast.

Stabilization Energy: The difference in energy between the delocalized structure and the conventional structure for a compound. The stabilization energy of 'benzene is 150 kilojoules per mole, which represents the difference in energy between a Kekule structure and thereal hybrid structure which, being of lower energy is more stable.

Stainless Steel: A class of chromium steels containing 70%–90% iron, 12%–20% chromium, 0.1%–0.7% carbon.

Stalloy: Steel containing 3.5% silicon, having low energy losses due to hysteresis. Used in portions of electrical apparatus that are subjected to alternating magnetic fields.

Standard Atmospheric Pressure: A pressure sufficient to support a column of mercury 760 mm in height. This is the average pressure exerted by the atmosphere at sea level. Gas pressures are thus generally given in unit of mm Hg or in units of atmosphere. One atmosphere (1 atm.) corresponds to 760 min Hg. A mm Hg pressure is referred to as a torr.

Standard Electrode: A half cell used for measuring electrode potentials. The hydrogen electrode is the basic standard but, in practice, coloured electrodes are usually used.

Standard Solution: The solution whose concentration is known.

Standard State: The condition in which a substance is in its most stable form at 25°C and 1 atm.

Standard Temperature: An internally agreed value for which many measurements are quoted. It is the melting temperature of water 0°C

Stannane: A colourless poisonous gas prepared by the action of lithium aluminium hydride on tin (II) chloride.

Standing Wave: A wave whose peaks and node do not change position.

Stannic Bromide ($SnBr_4$): A white crystalline solid obtained by heating tin in bromine vapour. It fumes strongly in moist air and soluble incold water.

Stannic Fluoride (SnF_4): A white crystalline substance obtained by dissolving stannic hydroxide in hydrofluoric acid. It is better obtained by the action of anhydrous hydrofluoric acid on stannic chloride.

$$SnCl_4 + 4HF \rightarrow SnF_4 + 4HCl.$$

It is hydroscopic in nature. It absorbs ammonia forming a double compound, SnF_4–NH_3.

Stannic Iodide (SnI_4): A red octahedral crystals obtained beating tin with iodine. It is soluble m organic solvents

Stannous Bromide ($SnBr_2$): A light yellow crystalline compound obtained by heating tin with hydrobromic acid or by distilling tin with mercuric bromide.

Starch: A polysaccharide that series as storage reservoir of plants; occurs almost all plants and is a dietary component of animals. Starches are extracted commercially from barley, maize, potato, rice, sorghum, tapioca and wheat. Chemically, starch is not a single molecule but a mixture of 10–20% amylose and 80–90% amylopectin.

Starting Point Line: Position on chromatogram where the sample is applied. Usually 10–20 mm from the bottom of the plate or paper strip.

Statcoulomb A unit of electric charge in metric system, equal to $1/3 \times 10^9$ coulomb, i.e., 1 coulomb = 3×10^9 statcoulombs.

States of Matter: The three physical conditions in which substances exist: solid, liquid and gas.

Stationary Phase: The immobile phase bound tothe solid support, e.g., surface of an adsorbent in TLC, the liquidsupport coated on fine particles of silica or alumina (in GC and HPLC) the water molecules embedded in cellulose ions forming part of an ion exchange resin etc.

Steam Distillation: A technique of purifying substances by making use of Dalton's Law of partial pressures to lower die boiling point of the mixture. When two immiscible liquids are distilled, the boiling point will be lower than that of the more volatile component and consequently will be below 100°C if one component is water. A steam is bubbled through an impure liquid in flask heated on sand bath. The vapours of the substance mixed with steam are condensed in water condenser. The two being immiscible with waterare separated in a separating funnel. The method is partyicularly useful for recovering materials from tarry mixtures.

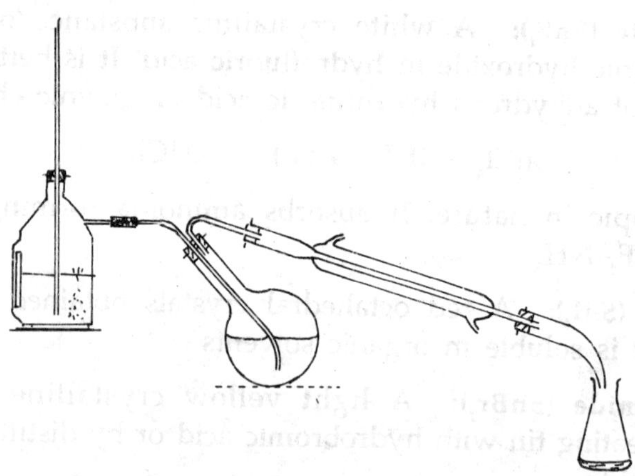

Steam distillation

Stearic Effect: An effect in which the shape of the molecule influences its reactions. A particular example occurs in molecules containing large groups, which hinder the approach of a reactant.

Steel: An alloy of iron with carbon, and other elements like manganese, phosphorus and silicon, The carbon content of steel generally lies between 0.1% and 1.5%. It is used for many purposes by alloying metals like nickel, manganese, chromium, tungsten etc. It is generally hard and wearproof.

Steradian (Sr): One of the two SI units, supplementary unit of solid angle. The surface of a sphere subtends a solid angle of 4p at its centre. The solid angle of a cone is the area intercepted by the cone on the surface of a sphere of unit radius.

Steroid: A generic term for substance having nuclear carbon skeleton of the sterols or a very similar stricture.

Still: An apparatus for distillation, especially for distilling alcohol or petroleum, also the vessel in which a mixture is heated for distillation.

Stoichiometric: The proportions in which elements from compounds. A stoichiometric compound is one in which the atoms have combined in small whole numbers.

Stomach Poison: The chemicals that are ingested by the insect and research its alimentary canal to kill it. Because they are mostly applied prior to insect attack or insect resurgence, they are also called protective insecticides.

Streptomycin ($C_{21}N_7O_{12}$): An antibiotic substance produced by the Streptomyces fungus. It is effective against several types of disease bacteria, including some against which penicilin is inactive.

Strontium: A silver metal which occurs in nature as celesite, $SrCO_3$ and strontianite $SrCO_3$. It belongs to alkaline earth metal. It is obtained by the electrolysis of fused strontium chloride. It reacts directly with oxygen, nitrogen, hydrogen, sulphur and halogens. Its compounds are used in fire works and in refining of sugar.

Strontium Carbonate (Sr-CO_3): A white soft powder which occurs in nature as strontianite. It can be obtained commercially from celestine by fusion with sodium carbonate.

$$SrSO_4 + Na_2CO_3 \rightarrow Na_2SO_4 + SrCO_3$$

It resembles calcium carbonate but decom poses on heating not so readilyas calcium carbonate. It is used for preparing strontium hydroxide.

Strontium Chloride ($SrC1_2$): A hexahydrate crystal obtained by dissolving strontium oxide, hydroxide or carbonate in dilute hydrochloric acid. It is very soluble in water but less deliquescent than calcium chloride

Strontium Hydroxide [Sr-$(OH)_2$]: A solid which occurs as the octahydrate [Sr(OH_2.–$8H_2O$]. It is obtained by heating strontium carbonate in steam

$$SrCO_3 + H_2O \rightarrow Sr(OH)_2 + CO_2$$

It resembles slaked lime, but is more soluble. Its aqueous solution is strongly alkaline. It is used for refining of sugar.

Strontium Nitrate [$Sr-(NO_3)_2$]: A tetrahydrate crystal obtained by dissolving strongium oxide, hydroxide or carbonate in dilute nitricacid. The large preparation is done by the interaction of strontium chloride and sodium nitrate. It is soluble in water though less than calcium nitrate.

Strontium Oxide (SrO): A white powder obtained by heating the carbonate or nitrate. It slakes on addition of water just like quicklime to form $Sr(OH)_2$which is used in the recovery of sugar.

Strontium Sulphate ($Sr–SO_4$): A white solid occurs in nature as celestine. It is obtained adding a soluble sulphate to any strontium salt solution as it is insoluble in water. The solubility is less than of calcium sulphate but greater than that of barium sulphate. It is used in the preparation of other strontium compounds.

Structural Formula: A formula that shows not only the number and kind of atoms in a molecule, but also the arrangement of the atoms.

Su: Strontium unit; a measure of strontium-90 and hence calcium in an organic sample. 1 Su = 10^{-12} curies of ^{90}Sr, equivalent to one gram of calcium in the sample.

Suberic Acid [$HOOC–(CH_2)_6COOH$]: A white crystalline dibasic acid, obtained from castor oil and used in the manufacture of plastics and plasticizers.

Sublimation: The process in which a solid goes directly into a vapour without going through the liquid state. This process is used to purify organic compounds like camphor, naphthalene etc.

Sub-micron: A term used to describe particles, in the size range 10 mm - 1mm which are invisible in an ordinary microscope but quite enough to appear in an ultramicroscope.

Subshell: A concept used in the Bohr theory of atomic structure. Each electron shell is divided into sub-shells, for which all the electrons have the same azimuthal quantum number. The shells are designated by the letter *s, p, d, f.*

Substituent: An atom or group that replaces another atom or group in a molecule during substitution reaction, e,g., in nitration of benzene to nitrobenzene the substituent is a nitro group (NO_2) that substitutes for hydrogen.

Substitution Reaction: A reaction in which an atom or group of atoms in an organic molecule is replaced by another atom or group, e.g.,substitution of an hydrogen atom in an alkane for halogen to form ahaloalkane. Substitution reactions are of three types: (1) nucleophille substitution, e.g,, attack is by a nueleophile as hydrolysis of chloroethane to produce ethanol in presence of alkali $C_2H_5Cl + OH^-$ ® $C_2H_5OH + Cl^-$ (2) electrophilic substitution in which the attacking substituent is an electrophile that attacks an electron rich site in a molecule, e.g., nitration of benzene to get nitrobenzene (3) free radical substitution where a free radical is the attacking substituent, e.g., halogenation of an alkane to form haloalkane.

Substrate: A substance whose reactivity is increased by a specific enzyme.

Subtractive Process: The process of producing colours by mixing three different dyes or pigments together. The final colour is produced by the absorption of different wavelengths of light.

Sucrose or Cane Sugar or Beet Sugar ($C_{12}H_{22}OH$): A colourless crystalline substance, sweet in taste and very soluble in water. It is optical active compound having specific rotation = + 66.5°

Sugar: Sugars are sweet-tasting simple carbohydrates. sugars have molecules consisting of a chain of carbon atom with—OH groups attatched, and either an aldehyde or ketone group. They can exist in a chain form or in a ring formed formed by the reaction of the ketone or aldehyde group with an OH group. Monosaccarides are simple sugars are that cannot be hydrlysed to to sugars with fewer carbon atoms. Two or more monosaccharide units can linked in disaccharides, trisaccharides, etc.

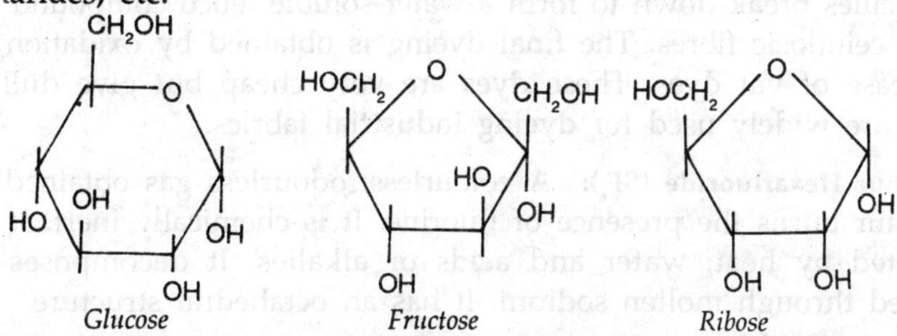

Examples of monosaccharides

Sulphadiazine: Obtained by using 2-amino-pyrimidine in place of 2-aminopyridine. It is widely used as a cure for mild infections.

Sulphanilic Acid or p-Aminobenzene Sulphonic Acid ($p\text{-}NH_2C_6H_4SO_3H$): A white solid which before melting point. It is very soluble in water but insoluble in organic solvents. When treated with concentrated nitric acid, it gives p-nitroaniline and on bromination the sulphonic group is replaced by Bratom to give striromoaniline.It is used in the manufacture of dyes and sulphanilamide drugs.

Sulphide: (1) Binary compound of sulphur with a more electro positive element, e.g., sodium sulphide Na_2S. Sulphides contain S^{2-} anions but similar compounds containing a chain of sulphur atoms can also be formed. Such compounds are called polysulphides. (2) Dialkyl or diaryl derivatives of sulphur; generally known as thioethers.

Sulphur: A low melting nonmetallic solid which exists in nature as combined and active states, It is extracted mainly by Frasch process. It belongs to group VI of the periodic table. It is used in the manufacture of sulphur dioxide, sulphuric acid, sulphite acid, carbon disulphide. It is used as disinfectant, fungicide, insecticide and as medicine.

Sulphur Dichloride Dioxide (Thionyl) Chloride, ($SOCl_2$): A colourless liquid b.p. 78°C. It fumes in moist air due to the hydrolysis

$$SOCl_2 + H_2O \rightarrow 2HCl + SO_2$$

It is used as chlorinating agent.

Sulphur Dioxide (SO_2): A colourless gas with a choking penetrating smell; liquid SO_2 is used in bleaching, fumigating, and as are frigerant.

Sulphur Dyes: Dyes made by heating certain organic substances with sulphur and sulphides. They are usually of polymeric nature, and are insoluble in water, but when heated with sodium sulphide the large molecules break down to form a water-soluble leuco compound which dyes cellulosic fibres. The final dyeing is obtained by oxidation, as in the case of vat dyes. These dyes are very cheap but give dull hues; they are widely used for dyeing industrial fabrics.

Sulphur Hexafluoride (SF_6): A colourless, odourless gas obtained when sulphur burns the presence of fluorine. It is chemically inert. It is not affected by heat, water and acids or alkalies. It decomposes when passed through molten sodium. It has an octahedral structure.

Sulphuric (IV) Acid (Sulphurous Acid, H_2SO_3): A dibasic weak acid obtained by passing sulphur (VI) oxide into water. It is a reducing agent but oxidizes strong reducing agent like hydrogen iodide

$$H_2SO_3 + 4HI \rightarrow S\downarrow + 3H_2O + 2I_2$$

It also reduces hydrogen sulphide to sulphur and is itself reduced to sulphur.

Sulphuric acid Sulphuric (V) Acid (H_2SO_4): A colourless oily liquid obtained industrially by contact process. It acts as an oxidising agent and also as a dehydrating agent. It is used in the manufacture of fertilizers, rayons, detergents and in batteries.

Sulphur (VI) Oxide (Sulphur Trioxide, SO_3): A fuming white solid obtained by heating ferrous sulphate or when a mixture of sulphurdioxide and oxygen is passed over heated spongy platinum or V_2O_5. It dissolves in concentrated sulphuric acid to give oleum. It is used in the manufacture of sulphuric acid, and for drying gases. It has a plane triangular structure.

Sulphur Point: The temperature of equilibrium between liquid sulphur and its vapour at a pressure of one standard atmosphere 444°C.

Sulphuryl Chloride (SO_2Cl_2): A colourless fuming liquid obtained by the direct combination of sulphur dioxide with liquid chlorine in the presence of camphor. $SO_2 + Cl_2$ ® SO_2Cl_2. It is used for the industrial preparation of the acid chlorides of organic acids for their alkali salts.

Super Heating: The raising of a liquid's temperature above its boiling temperature by increasing the pressure.

Supernatent: Denoting a clear liquid that lies above a sediment or a precipitate.

Supercooled Liquid: A liquid at a temperature below its freezing point.

Supercritical Fluid: A substance at a temperature above its critical temperature.

Super Oxide: An organic compound containing the O_2 ion.

Super Phosphate: A mixture—mainly calcium hydrogen phosphate and calcium sulphate used as a fertilizer. It is made from calcium phosphate and sulphuric acid.

Supplementary Units: The dimension less units the radian and the steradian used along with base units to form divided units.

Surface Energy: The surface energy of a system is defined as the work necessary to increase the surface by unit area against the force of surface tension.

Surface Tension: A measure of inward force on the surface of liquid, the force which must be overcome to expand the surface area. It decreases with increasing temperature.

Suspension: A system in which small particles of a solid or liquid are dispersed in a liquid or gas.

Sweating: The process of removing entrapped oil and lowmelting waxes from the filter cake formed in dewaxing processes.

Synergists, 1. 2, 3-Benzthiadiazole: Substituted 1, 2, 3-benzthiadiazole derivatives which inhibit phenolase enzymes and act as synergists for pyrethrins.

Synergists Analogue: Compounds that resemble in structure to a particular insecticide and synergise its activity, e.g., 4, 4-dichlorophenyl methyl carbinol (DMC) is a synergist for DDT and salithion is a synergist for malathion.

Synergists, Benzo 1, 3-Dioxole: Pyrethrum synergists derived from methylene-dioxyphenyl benzene, e.g., piperonyl butoxide (PBO), sulfoxide, propyl isomer etc. This group of compounds inhibit mixed function oxidase and synergize, besides pyrethrins, organophosphorus, organocarbamates and organochlorine insecticides also.

Synergists, Phenyl 2-Propynyl Ethers: 2-Propynyl ethers of substituted phenols which synergise N-methyl carbamate insecticides. The mostactive compound of the series is 3-(2-nitro-4-chlorophenoxy) proplyne.

Synthesis Gas: A mixture of carbon monoxide and hydrogen produced by steam reforming of natural gas.

$$CH_4 + H_2O \rightarrow CO + 3H_2$$

Synthesis gas is a useful starting material for manufacture of a number of organic compounds.

T

Talc (Hydrated Magnesium Silicate, $3MgO.4SiO_2H_2O$) Used as a lubricant and in talcum powder.

Tall Oil A resinous substance obtained as a byproduct in the manufacture of wood-pulp, used in soaps and paints.

Tallow The rendered fat of animals, particularly cattle and sheep. It consists of various glycerides.

Tannic Acid: A white amorphous solid, extracted from gall-nuts.

Tanning: The conversion of raw animal hide into leather by the action of substance containing tannin, acid, or other agents.

Tannins: A class of complex organic compounds of vegetable origin. Compounds consist of mixtures of derivatives of polyhydroxy benzoic acids.

Tantalum: A grey transition metal. It is found as $Fe(TaO_3)_2.Fe(NbO_3)_2$. It is stable in air and unites with nonmetals on heating. It is used in making metal plates, screw and wires needed for repairing fractured bones. It is also used in turbine blades.

Tar: A thick, black, sticky liquid obtained during the destructive distillation of coal.

Tartaric Acid, a-dihydroxy succinic acid HOOC-.CHOH. CHOH.COOH): A colourless crystalline substance synthesised by boiling dibromosuccinic acid with silver oxide suspended in water. It is soluble in water and alcohol but soluble in ether. It is optically active. Potassium acid tartrate is used in dyeing and making baking powder. Potassium antimony tartate is used in dyeing in calicoprinting and in Fehling's solution and in silvering of mirrors.

Tau Particle: A heavy lepton that reacts liv the weak interaction. It has a very short life time (about 5 x 10^{-12} second) and a mass of approximately 1800 MeV. There is strong evidence for its existence.

Taurine ($NH_2(CH_2)2SO_3H$): A white crystalline substance, m.p. 328°C, obtained from the bile of mammals.

Tautomerism: The type of isomerism in which a compound exists simultaneously in two forms in equilibrium with each other.

Technetium: A radioactive transition element produced by irradiation of molybdenum oxide with neutrons.

Tellurium: It exists in two forms, a crystalline and amorphous form. The crystalline tellurium is steel grey, metallic looking, brittle solid. It burns in air on beating to give tellurium dioxide (TeO_2). It is added to lead to increase its resistance to chemical attack, alloy resist bursting. It is also used in electroplating of silverware, when a dark finish is needed.

Tellurium Hexafluoride ($Te\text{-}F_6$): A gas formed by direct union of the elements SF_6. It is completely hydrolysed by water.

$$TeF_6 + 6H_2O \rightarrow H_6TeO_6 + 6HF.$$

Tellurium Monoxide (TeO): A black solid obtained on heating tellurium sulphate.

$$TeSO_3 \rightarrow TeO + SO_2$$

Tellurium Trioxide (TeO_3): An orange yellow powder obtained by heating telluric acid to redness. It is insoluble in water. On heating it dissociates into TeO_2 and O_2.

Telomerization: Polymerization to a very low mol. wt. polymer or condensation involving a few molecules of olefine.

Tempering: The method of reheatting a quenched steel to relieve internal stresses, reduce hardness and tensile strength and restore dueffity and resistance to impact loading. It is carried out within the temperature range 200-500° for plain carbon steels.

Temporary Hardness: A kind of water hardness due to the presence of calcium, iron and magnesium bicarbonates.

Teratogen: A chemical which gives malfunction generally in the form of mutations or tumors.

Terbium: A soft ductile malleable silvery element of the lanthanoid sedes of metals.

Ternary Compound: A chemical compound formed from three elements, e.g., Na_2SO_4 or $LiAlH_4$.

Terpenes: In the strict sense terpenes are volatile, aromatic hydrocarbons of the empirical formula $C_{10}H_{16}$.

Tetra Chloro Methane (Carbon Tetrachloride, CCl_4): A colourless liquid; b.p. 77°C, which has sickly smell. Its vapours are noninflammable. It is used as a solvent for fats, oils, resins. It is also used as fire extinguisher under the name of pyrene.

7, 7, 8, 8-Tetracyanoquinone Diniethane (TC-NO): Accepts an electron from suitable donors forming a radical anions. It belongs to a class of compounds called II acids. It is used for colorimetric determination of free radical precursors, electrolytic capacitors and ion specific electrodes.

Tetraethyllead, [(C_2H_5)4Pb]: A colourless liquid, insoluble in water but soluble in ether. It is highly poisonous and needs special precautions in handing. It is used for raising octane number of petrol.

Tetrahedral Compound: A compound in which an atom has four bonds directed towards the corner of regular tetrahedron.

Tetrahydrate: A crystalline hydrated compound having four molecules of crystallization per molecule of compound.

Tetrahydrate: A crystalline hydrated compound having four molecules of crystallization per molecule of compound,

Tetra Hydrofuran, (C_4Hg-O): A colourless liquid widely used as a solvent.

Tetranitromethane, [C(N–O_2)$_4$]: A liquid, b.p, 1,26°C. It is obtained by nitration of ethanoic anhydride with anhydrous nitric acid. Decomposed by alcoholic solution of KOH to nitroform, It is used for testing unsaturation.

Thalium: A soft malleable ductile silvery element of the lanthanoid series of metals.

Thallic Sulphate [$T1_2(SO_4)_3$–.7H_2O]: Formed by dissolving thallic oxide in dilute sulphuric acid. It is decomposed in aqueous solution and a basic salt Te(OH) SO_4.-2H_2O is formed with K_2SO_4. It forms K_2SO_4.-$Tl_2(SO_4)_3$.-8H_2O which is not a true alum

Thallium: A soft white metal occurs in nature associated with pyrites or zinc ore. It belongs to IIIA group. It oxidises in moist air and does not readily dissolve sulphuric and nitric acid. It is used in the production of a very refractive, in photocells and in infrared detectors.

Thallous Carbonate (Tl_2CO_3): Obtained by the adsorption of CO_2 by thallous hydroxide solution.

Thallous Chloride (TlCl): Precipitated when hydrochloric acid is added to a solution of thallous sulphate. It becomes violet in exposure to light. It is sparingly soluble in ammonia.

Thebaine ($C_{19}H_{21}NO_3$): A white insoluble substance, m.p. 193°C, present in Optium in small quantities.

Theobromine ($C_7H_8N_4O_2$): A white insoluble crystalline alkaloid, m.p. 337°C, that is isomeric which theophylline, m.p. 272°C.

Thermal Pollution: The heat lost to our environment through our various energy conservation.

Thermite (Thermit): A mixture of Iron (III) oxide (3 part) and aluminium powder (1 part). When the magnesium ribbon fuse is lighted, the ignition mixture catches fire and reduction of iron (III) oxide by aluminium powder occurs.

$$Cr_2O_3 + 2AI \rightarrow 2Al_2O_3 + 2\ Fe$$

Thermobalance: A balance in which the sample can be heated to allow thermal analysis.

Thermochemistry: The branch of chemistry which deals with heat of reaction, solvation etc.

Thermodynamics: The quantitative treatment of the relationship between different forms of energy. The first law of thermodynamics states that energy can neither be created nor destroyed even though it may be changed from one form to another. The second law of thermodynamics is based on the fact that it is impossible to convert heat into an equivalent amount of work without producing other changes in some parts of the system. The third law of thermodynamics states that the entropy of a substance tends to zero as its thermodynamic temperature approaches zero.

Thermoplastic: A polymerized substance which forms plastic on heating and can be shaped by pressure and heat, without changing the chemical properties.

Thermosetting Plastic: A polymerized substance which first forms plastic on heating and can be shaped by pressure and heat without change. A different substance is obtained which cannot be converted to plastic by heating.

Thiazole (S.CH:N.CH:CH): A collourless volatile liquid, b.p. 116.8°C, whose molecule consists of a fivemembered ring. Derivatives are used in dyestuffs and in medicine.

Thin-layer Chromatography: A form of chromatography in which the stationary phase consists of a thin layer of alumina slurry of a glassplate. After selective absorption of the mobile phase the plate is dried in an oven. The technique is very similar to that of paper chromatography.

Thiocarbamide or Thiourea [$(NH_2)_2CS$]: A white crystalline solid, fairly soluble in water and alcohol. It is obtained by heating ammonium thiocyanate at 443 K.

$$NH_4CNS \rightarrow (NH_2)_2\,C = S.$$

It is used in the manufacture of plastics and as a reagent in the estimation of cadmium and lead.

Thioacetamide [$CH_3CS–NH_2$]: Colourless soluble crystalline substance, m.p. 115-16°C, used as a source of hydrogen sulphide.

$$NH_4CNS \rightarrow (NH_2)_2\,C = S$$

It is used in the manufacture of plastics and as a reagent in the estimation of cadmium and lead.

Thiochrome (Cl_2HO_4OS): A yellow prism. It is soluble in methanol fairly soluble in water. It is obtained by oxidizing thiamine.

Thio Alcohol: A compound of the type RSH, similar to an alcohol will the oxygen atom replaced by sulphur.

Thiocyanate: A salt or ester of thiocyanic acid.

Thiocyanic Acid (HSCN): An unstable acid that forms salts called thiocyanates.

Thio Ethers: A group of compounds with the general formula RSW where R and R′ are hydrocarbon radicals.

Thiols: A class of organic compounds for the general formula RSH, with sulphur attached directly to carbon; they are the sulphur analogues of alcohols, containing SH instead of OH group. Formerly called mercaptans.

Thionin ($Cl_2H_9N_3S$): A dark brown thiazine derivative, used as a dye in microscopy.

Thionyl Chloride ($SOCl_2$): A colourless pungent smelling liquid obtained by the action of phosphorus chloride on sulphur dioxide or sodium sulphite.

$$SO_2+PCl_5 \rightarrow SOCl_2+POCl_3.$$

In contact with water it is decomposed at once forming sulphurous acid and hydrochloric acid.

$SOCl_2 + 2H_2O$ ® $H_2SO_3 + 2HCl$. It is used as a reagent for replacing the hydroxyl group in organic compounds by chlorine.

Thiosulphate: A salt or caster of thiosulphonic acid. Inorganic thiosulphates contain $S_2O_3^{2-}$ ion e.g., sodium thiosulphate $Na_2S_2O_3$.

Thiosulphuric Acid ($H_2S_2O_3$): An unstable acid formed by replacing one oxygen atom of sulphuric acid (H_2SO_4) by one sulphur atom. It is known only in solution or in the form of its salts or esters, the thiosulphates.

Thiphene or Thiacyclopenta-2, 4 diene (C4H4S): A colourless liquid smelling like benzene obtained by passing a mixture of acetylene and hydrogen sulphide through a tube containing Al_2O_3 at 670K.

Thiouracil ($C_4N_4N_2OS$): A white powder obtained by condensing sodium ethylforrnylethanoate with thiourea. It decreases the activity of the thyroid gland.

Thixotropy: The change of viscosity with time. Fluids that suffer a change in viscosity on standing are said to be thixotropic.

Thorium: A silvery white radioactive soft metal. It resembles lanthanidemetals in chemical reactivity. It gets tarnished when exposed to air. The important minerals are thoritc, orangeite and nionazite. Its oxide (ThO_2) is used in the preparation of incandescent gas metals.

Thyroxine: A hormone secreted by thyroid gland and found in the protein thyroglobulin. Chemically it is b-3, 5-dilodol (3, 5-diiodo-4-hydroxyphenoxy) phenyl-a-aminopropionic acid.

Tin: A silver white metal and is softer than zinc but harder than lead. The main ores is cassiterite or tin stone, SnO_2. The fourth element of IVA group. It is malleable and ductile at 370 K. A cracking sound is heard when a rod of tin is bent due to the crystals rubbing against one another. It is weakly electronegative and relatively less active. Air and water do not attack tin at ordinary temperatures. It reacts with acids, alkalies. Its compounds are used as mordants in dyeing and calicoprinting. It is used in tinning utensils and for preparing tin foil.

Tin (II) Chloride (Stannic Chloride, $SnCl_2$): A transparent solid obtained by dissolving tin in hydrochloric acid. It is soluble in water but the basic chloride is precipitated on standing due to hydrolysis. Sn^{2+} ions act as an electron donor (reducing agent). It is used as mordant indyeing and as reducing agent in the laboratory.

Tin (VI) Chloride (Stannic Chloride, $SnCl_4$): A colourless fuming liquid obtained by passing chlorine into molten tin kept in a retort. It is highly reactive substance and can catch fire with turpentine.Its pentahydrate is used as mordant under the name butter of tin and also used for preparing fire-proofing cotton.

Tin (II) Oxide (Stannous Oxide, SnO): A dark grey or black powder which burns in air to give stannic oxide. It is an amphoteric oxide and dissolves in acid and alkali.

Tin (V1) Oxide (Stannic Oxide, SnO_2): A white powder obtained when tin is burnt in air, It is used in getting white glaze for tiles and milky glass articles and electric light shades.

Tin (II) Sulphide (Stannous Sulphide, SnS): A grey crystalline, compound obtained by burning fin in sulphur vapour. It on heating above 265°C turns into tin (IV) sulphide.

Tin (IV) Sulphide (Stannic Sulphide, SnS2): A yellowish solid obtained by passing hydrogen sulphide through an acidified stannic salt solution. It is soluble in yellow ammonium sulphide due to the formation of ammonium thiostannate.

$$SnS_2 + (NH_4)_2S \rightarrow (NH_4)_2SriS_3$$

It is used a pigment for bronzing articles of wood.

Tin Hydride (SnH4): Obtained by the action of hydrochloric acid on analloy of tin and magnesium.

Tinning: A method of covering the utensils with a thin layer of tin.

Tischenko Reaction: On heating benzaldehyde with aluminium alkoxide and a little anhydrous $AlCl_3$ or $ZnCl_2$, it undergoes an intermolecular oxidation and reduction similar to cannizaros reaction but the alcohol and acid instead of appear in such react to produce benzyl benzoate.

$$Al(OC_2H_5)_3$$

$$2C_6H_5CHO \rightarrow Al(OC_2H_5)_3\ C_6H_5CH_2OOC\ C_6H_5$$

Titanium: steel grey transition metal. It is light, ductile, has a hightensile strength and a low coefficient of expansion. The most abundant ore of titanium is the titaniferous iron stone, called ilmenite ($FeTiO_3$) and other one is rutile, TiO_2. It is used to construct ships, marine machinery. Jet aircraft, as a rack material in the anodization of the metal and ass alloying agent in the manufacture of ferrous and nonferrous alloys.

Titanium (II) Chloride ($TiCl_2$): Due to its strong tendency to get oxidized to TiCL, it behaves as a strong reducing agent. It is an ionic compound. It undergoes disproportionation to give $TiCl_2$ and Ti:

$$Cl_2 \text{ and } Ti_2TiCl_2\ TiC_4 + Ti$$

Titanium (III) Chloride ($TiCl_3$): A violet powder obtained by passing vapours of titanium (IV) chloride mixed with excess of hydrogen through a red hot tube heated to about 920 K.

Titanium (IV) Chloride. (Titanium Tetrachloride, TiC4): A colourless strongly fuming liquid, b.p. 136.4°C and m.p. –30°C. It is soluble in water but is hydrolysed by excess depositing TiO_2. It is used in the preparation of pure titanium.

Titanium (IV) Oxide (Titanium Dioxide, TiO_2): A white powder insoluble in water. It slowly reacts with acid, but easily dissolves in fused alkalies forming Titanates (TiO_3^{2-}). It is used as a pigment.

Titration: A volumetric technique in which a solution of known concentration is added to a solution of unknown concentration until end point is obtained.

Tollens Reagent: An ammoniacal solution of silver oxide which is used as a test for aldehydes., which, unlike ketones, cause did deposition a silver mirror.

Tonne (Symbol T): A unit of mass equal to 10^3 kilograms.

Tracer: An isotope of an element which is used to investigate chemical reaction or physical process.

Tranquilizer: A drug used to reduce tension and anxiety, without impairing alertness or physical processes.

Trans: A geometric arrangement refers to one with like group opposite each other.

Transition Elements: A group of elements having partially filled d-subshells in any of their common oxidation number. Scandium is the first of a series of nine elements in the periodic system in which the number of electrons in die 3d sub-shell of the atoms is increasing from one to ten. The second series begins with yttrium and ends with silver. The third series begins with lanthanium and ends with gold. A fourth such series starts with actinium.

The transition elements are metallic in nature, and most of them have high melting points, high boiling points, high densities, high tensile strength and high heat of sublimation. They are good conductor of electricity. Many compounds of this series are coloured and show magnetic moment.

Transition Temperature: A temperature at which some definite physical changes take place in a substance.

Transmutation: A conversion of one element into another by virtue of radioactive decay or by bombardment of the nuclei with particles.

Transport Number (Symbol T): In an electrolyte, the transport number of an ion is the fraction of the total charge carried by that type of ion in conduction.

Transuranic Elements: The elements of the actinoid series which have higher atomic numbers than uranium.

Triatomic: A molecule, radical or ion which consists of three atoms, e.g., O_3 and SO_2.

Tribasic Acid: An acid having three atoms of acidic hydrogen per molecule, e.g., phosphoric acid, H_3PO_4.

Tribromoethanol (CBr_3 CH_2OH): A white crystalline powder, m.p. 7982°C, used as a veterinary anaesthetic.

Tribromo Methane (Bromoform, $CHBr_3$): A colourless liquid, b.p. 149.5 and smells like chloroform, which closely resembles chemically. It is prepared industrially by the electrolysis of an aqueous solution facetone, or ethanol containing sodium carbonate and potassium bromide.

Trichloroacetic Acid (CO_3COOH): A colourless, deliquescent solid and one of the strongest organic acids. It gives chloroform when boiled with dilute alkali or even water.

Trichloro Ethanol (Chloral, CO_3CHO): A colourless liquid, b.p. 87°C. It is soluble in water, ethanol and ether. It is used as a sedative.

Trichloroethylene ($CHCl: CCl_2$): A colourless liquid, b.p. 87°C; widely used as industrial solvent, in dry cleaning, and as an anaesthetic.

Trichloro Methane (Chloroform, $CHCl_3$): A colourless liquid obtained by heating ethanol or acetone with bleaching powder. It is used as anaesthetic and also in production of chloretone, which is used as drug.

Triethylamine [$(C_2H_5)_3N$]: A colourless inflammable liquid, b.p. 89-90°C, used as a solvent.

Triglyceride: A glyceride in which esters are formed with all three —OH groups.

Trihydrate: A crystalline hydrated compound that contains three molecules of water of crystallization per molecule compound.

Tri-iodo Methane (Iodoform, CHI_3): A yellow crystalline solid, m.p.119°C. It is insoluble in water but is readily soluble in ethanol. It is used as antiseptic.

Triol: An alcohol that has three hydroxyl (–OH) groups (on three different carbon atoms of the chain) per molecule of the compound, e.g., glycemol, $CH_2(OH)–CH(OH–CH_2$ (OH) (at. trihydric alcohol).

Trimer: A molecule formed by addition of three identical molecules.

Trimethyl Amine [$(CH_3)_3N$]: A colourless gas. It is used as a source of methyl chloride.

1, 3, 5-Trinitrobenzene or S-Trinitrobenzene [$C_6H_3(NO_2)_3$]: A colourles solid (m.p. 395 K). It is obtained by nitration of m-dinitrobenzene with a mixture of fuming nitric acid and, fuming sulphuric acid, It has a greater explosive power and is used in preparing explosives.

Triol: An alcohol that has three hydroxyl groups (–OH) per molecule of compound.

Triphenylamine [$(C_6H_5)_3N$]: A colourless crystalline solid (m.p. 400 K). It is two weak a base to form salts with acid. It gives a blue solution when dissolved in concentrated sulphuric acid.

Triple Bond: A covalent bond obtained between two atoms in which three pairs of electrons are shared, e.g., in nitrogen molecule (N_2)

$$N \equiv N$$

Triple Point: The only point at which the gas, liquid and solid phases of a substance can coexist in equilibrium.

Triplumbic Tetraoxide (Red Lead, Pb_3C_4): A fiery red powder obtained by heating litharge (PbO) at about 450°C in air.

$$6PbO + O_2 \rightarrow 2Pb_3O_4$$

It is insoluble, in water but turns darker and darker on heating and decomposes to liberate oxygen gas.

$$2Pb_3O_4 \rightarrow 6PbO + O_2$$

It forms brown precipitate with dilute nitric acid.

$$Pb_3O_4 + 4HNO_3 \rightarrow 2Pb(NO_3)_2 + PbO_2 + 2H_2O$$

It is considered to be a orthoplumbate, Pb_2pbO_4. It is mainly used in glass industry, for making pottery glaze, as pigment for oil paints and as an oxidizing agent in the laboratory.

Trisodium Phosphate (V) (Sodium Orthophosphate, Na_3PO_4): A hydratecrystal ($Na_3PO_4.12H_2O$) obtained by combining disodium hydrogenphosphate and excess of caustic soda and hot water.

$$Na_2HPO_4 + NaOH \rightarrow Na_3PO_4 + H_2O$$

It is readily soluble in water and is partially hydrolysed to give an alkaline solution. It is used to remove hardness from water.

Tritium (Symbol T): A radioactive isotope of hydrogen with a mass number 3.

Trival Name: A nomenclature for chemical substances which was used before chemistry was studied property, e.g., chalk, lime and blue vitrol.

Tropine ($C_8H_{15}NO$): A white crystalline hydroscopic soluble alkaloid m.p. 63°C.

Tropolones: 2-hydroxy-derivatives of cycloheptatrienone. The parent compound. tropolone forms colourless, needles m.p. 49-50°C.

Tropylium Ion: The positive ion $C_7H_7^+$, having a symmetrical seven membered ring of carbon atoms. The tropylium ion shows nonbenzenoid aromatic properties.

Trough: A flat container for liquid. The hollow between two waves.

Trouton's Rule: The latent heat of vapourization (DH_{vap}) of a liquid of molecular weight (M) and boiling point (TK) are related by expression.

Tschitschibaben Reaction: 2-aminoquinoline obtained on heating quninoline with sodamide in toluene solution.

Tungsten: A silvery white transition metal. The minerals of tungsten are wolframite, [Fe(Mn)WO_4]; scheelite $GaWO_4$ and cuproschectile, $CuWO_4$. It cannot he cast into rod or wire because of its high melting point. It is used in making filaments for electric lamps. It is also used to prepare special steel consisting of 14-20% tungsten and 33% chromium which is employed to make high speed lathe tools.

Tungsten (II) Chloride, Bromide and Iodide (WCl_2, WBr_2, WI2): WCl_2 and WBr_2 obtained by reducing corresponding pentahalides at higher temperature while WI2 is formed by union of the elements at red heat. They are strong reducing agents and liberates hydrogen from water.

Tungsten (IV) Bromide (WBr_4): Obtained by heating WCl_6 with liquid HBr under pressure. It is readily hydrolysed.

Tungsten (IV) Chloride (WCl_4): Obtained by the reduction of WCl_6 with hydrogen at high temperature. It is readily hydrolysed by water.

Tungsten Carbide (WC): A grey powder, m.p. 2780°C, obtained by direct combination of tungsten and carbon at 1600°C. It is almost as hard as diamond and is used in making abrasives and tools.

Tungsten Trioxide (WO_3): A yellow insoluble powder, m.p. 1473°C used in the manufacture of tungstates.

Tungstic Acid (H_2WO_4): Hydrated tungsten trioxide, A white crystalline powder that loses a molecule of water at 100°C, used in the manufacture of lamp filaments.

Tuyere: A pipe usually made of copper through which air is blown intoblast furnace.

Twinning: A term used in crystallography to denote a set of crystals that have one or more planes of faces in common but whose main developments away from that plane are not those of a single crystal.

Tyndall Effect: The scattering of light by colloidal particle, e.g., scattering of sunlight caused by dust particles.

Type Metal: An alloy of 60% lead 30% antimony, and 10% tin. Owing to the presence of antimony, it expands on solidifying and thus gives a sharp cast.

Tyrian Purple ($C_{16}H_8Br_2N_2O_2$): A purple Vat dye of great antiquity. It occurs in the shell fish.

Tyrosine: A white crystalline amino acid, m.p. 310°–320°C. Obtained from most proteins.

U

Ulman Reaction: The synthesis of diaryls by the condensation of aromatic halides with themselves or other aromatic halides, with tit concomitant removal of halogens by metal, e.g., copper powder thus bromobenzene gives di-phenyl.

Ultracentrifuge: A technique used to separate very small particles with the help of high speed centrifuge.

Ultramicroscope A device used to examine particles having colloidal dimensions.

Ultraviolet (uv): A kind of electromagnetic radiation, shorter in wavelength than visible light. The range is 1 rim to 400 nm.

Umber Natural $(Fe(OH)_3$ often containing MnO_2. A brown powder used as a pigment.

Unimolecular: A reaction in which only one molecule is involved.

Unit: A reference value of a quantity used to express other values of the same quantity.

Unit: Cell The smallest group of atoms, ion, or molecules that when repeated at regular interval in three dimensions will give die lattice of a crystal system.

Universal Indicator: A mixture of indicators, changes in its colour give an approximate value of pH.

Unsaturated Compound: An organic compound which contains double or triple bond between two of its carbon atoms.

Unsaturated Isobutylamides: A group of natural insecticides present in plants belonging to the families Composite and Rutaceae, e.g., pellitorine, piperlongumine, spilanthol, herculine, fagara-mide etc. Their insecticidal action is manifested by a rapid knock down effect similar to pyrethrins.

Uranium: A silvery white radioactive metal. It exists as U_3O_8 in pitchblende and carnotite, $KUO_2(VO_4)$. $3H_2O$. In India uranium occurs in monazite sand of Travancore. It forms a number of compounds in which it exhibits oxidation states of + 2, + 3, + 5 and + 6, It is used to get atomic energy by nuclear fission and in some alloys as source of transuranic elements. Ura-nium trioxide and sodium diuranate (Uranium yellow) are used as yellow pigments in glass and porcelain manufacture.

Uranium Dioxide: A black insoluble crystalline radio-active substance used as a fuel in gas-cooled reactors.

Uranium Hexaflouride (UF_6): A crystalline volatile compound obtained by heating uranium carbide or pentachloride with fluorine. It is deliquescent and fumes in moist air. It helps to separate uranium isotopes.

Uranium Penta Chloride (UCl_5): A deep red deliquescent solid obtained by prolonged chlorination of uranium oxides. It dissociates on heating into tetrachloride and chlorine.

$$2UCl_5 \rightarrow 2UCl_4 + Cl_2$$

It is soluble in water with hissing sound. It is soluble in many organic solvents.

Urea (Carbamide, NH_2 $CONH_2$): A white crystalline solid, m.p., 132°C soluble in water, ethanol but insoluble in ether. It is used as fertilizer and in the manufacture of urea formadehyde resin.

Urethane (Ethyl Carbamate, $NH_2COOC_2H_5$ Colourless Prisms, m.p 40–50°C by the action ammonia oil ethyl chloroformate. It used as an anaesthetic for laboratory animals.

Urethanes: Esters of carbamic acid. NH_2COOH Colour-less crystalline, solid, they distil without decomposi-tion. They are obtained by the action of ammonia on carbonic or chlorocarbonic ester or by treating alcohol, with urea nitrate or isocyanates.

V

Vacuum: A space containing gas below atmospheric pre-ssure. A perfect vacuum contains no matter at all.

Vacuum Distillation: The distillation of liquids done under a reduced pressure so that the boiling point is decreased.

Valence (Valency): The relative combining capacity of an atom or group of atoms compared with that of the standard hydrogen atom.

Valence Electron: The number of electrons present in the outer most orbit which can take part in forming chemical bonds.

Valeric Acid (Pentanoic Acid): A fatty acid which exists in several isomeric forms. Used in perfumes.

Vanadic Acid (HVO_3): A yellow insoluble crystalline substance. Other acids are formed by the addition of water molecules, e.g., H_3VO_4.

Vanadium: A steel grey transition metal, very hard and tenacious. The main ores are patronite, V_2S_5; carnotite, $K_2(UO_2)_2(VO_4).3H_2O$ and vanadipatronite $3Pb_3(VO_4)$. $PbCl_2$. It is used to form alloys with cobalt and iron. It is also used as a scavenger, strengthe-ning and toughing agent In iron, and steel industry.

Symbol V; a.n. 23; r.d. 6.1; m.p. 1920°C; b.p. 3388°C.

Vanadium (V) Oxide (Vana-dium Pentaoxide, V2O5): An orange red powder which melts without decomposi-tion at 660°C and recrystalizes on cooling to give yellow red needles. It readily dissolves in alkalies and acids. It is used in the manufacture of vanadium glass.

Van-Arkel Process: A method used to purify metals like titanium, hafnium and zirco-nium, by decomposing metal iodide on a hot filament, e.g., impure zirconium is converted to iodide with a limited amount of iodine.

$$Zr + 2I_2\ (g) \rightarrow ZrI_4\ (g)$$

The zirconium tetra iodide is decomposed upon contact with a hot filament.

$$ZiI_4\ (g) \rightarrow Zr\ (S) + 2I_2\ (g).$$

Vander Waal's Force: A weak intermolecular attraction due to the electrostatic attraction between the nuclei of one molecule and the electrons of the other, These forces have a short range and become significant only where the molecules are very close.

Vanillin or 4-Hydroxy-2-Meth-oxy Benzaldehyde: A pleasant smelling, colourless, crystal-line solid (m.p. 354 K). It gives blue coloration with ferric chloride solution. It is obtained from gluaiacol by means of Reimer Tiemann reaction.

It is used às flavouring and perfuming material.

Van't Hoff Isochore: The name given to the relation where K is the equilibrium constant of a reversible reaction, DH the enthalpy of reaction, T the absolute temperature and R die gas constant.

Van't Hoff Isother In, a reversible chemical reaction of the type

$$aA + bB = X + dD$$

.**Vapour Density (V.D.):** The ratio of the weight of a certain volume of the substance to the weight of the same volume of hydrogen, both volumes being measured identical conditions of temperature and pressure.

Vapour Pressure: The pressure caused by a vapour. It depends on the nature of the liquid or solid and the temperature.

Vat Dyes: The water insoluble coloured compounds which can be reduced to colourless (leuco) compounds which are soluble in alkali and readily reoxidised to the dye. Indigo is an outstanding example of Vat dye.

Venetian White: A mixture of white lead and barium sulphate, $BaSO_4$, in equal parts. It is used in paints.

Verdigris: A green deposit formed upon copper; it consists of basic copper carbonate or sulphate of variable composition.

Vetivone: ($C_{15}H_{22}O$) Two stereoisomeric ketones have been isolated from oil of vetiver.

Vicinal Position: Positions in a molecule at adjacent atoms for example the hydroxy groups in ethylene glycol, $CH_2(OH)CH_2(OH)$, are in vicinal position.

Victor Mayor: Method for Vapour Densities Method using the volume occupied by a known weight of vapour (generally measured as expelled air).

Vinasse: The residual liquid obtained after fermentation and distillation of beetroot molasses.

Vinyl Ether (C_4H_6O): A highly inflammable liquid (b.p. 301.3 K). It is obtained by the action of alkali on 2, 2-dichloroethyl ether. It is used as an anaesthetic.

Vinyl Fibres: Polymer derived from vinyl chloride and vinyl acetate were developed as plastics and resins but later the technique of emulsion copolymerisation was found applicable to the production of a variety of vinyl copolymers useful as fibres. Vinyon is copolymer of vinyl chloride (88%) and vinyl acetate (12%).

It is highly resistant to chemicals but its low softening point (338) limits its use as a textile fibre.

Vinylidene Chloride (CH_2: CCl_2): A colourless inflam-mable liquid, b.p. 32°C, that polymerizes to form polyvi-nylidene chloride.

Virgin Neutrons: Neutrons, produced by any means, before they have experienced a collision.

Virial Equation: A gas law that attempts to account for the behaviour of a real gas. It usually takes the form:

$$pv = RT + Bp + CP^2 + P^3 \ldots$$

Viscometer: An apparatus required to measure viscosity of the liquid.

Viscose: A thick tracely brownish liquid consisting mainly of a solution of cellulose xanthate in dilute sodium hydroxide.

Viscosity: The resistance to flow offered by a liquid is measured in terms of the property.

Vitamin: An organic substance required for regulating some of the body processes and preventing certain diseases.

Vitamin A ($C_{20}H_{29}OH$): Chemically it is a derivative of B-carotene. It is a fat soluble vitamin which is riot readily destroyed by cooking. It promotes growth in children and protects the body against certain germ diseases. It is found in milk, butter cream, cheese, eggs, vegetables and fruits.

Vitamin B Complex: A mixture of a number of water soluble vitamins named as vitamins B_1, B_2, B_6 and B_{12}. They help to regulate nervous system, Lack of these vitamins causes beri-beri skin eruptions, loss of appetite, nervous disorder and pellagra. It is found in milk, butter, whole grain, eggs, meat, green vegetables and brewer's yeast.

Vitamin B: Present in milk, green vegetables and rich polishings. Its deficiency causes beri-beri and loss of appetite.

Vitamin C or Ascorbic Acid (C_6H_8O): A water soluble vitamin present in tomatoes, leafy vegetables, green peppers, citrus fruits (lemons, oranges etc.) amla and papita. A deficiency causes structural defect of teeth and scurvy.

Vitamin D (Sunshine Vitamin): A complex containing at least four different compounds designated as D_1, D_2, D_3, D_4 etc. Ultra violet rays from sun change ergosterol, present in human skin. It is found in liver oil, milk and eggs.

Vitamin E: A fat soluble vitamin stable towards beat, present in animal and vegetable oils. Deficiency results in sterility.

Vitamin K: Two compounds designated as vitamin K_1 and K_2. It is present in leafy vegetables. Coagulating property found is due to the presence of vitamin K.

Vitreous: Resembling glass or having the structure of a gas.

Volatile: Easily converted into a vapour.

Volatility: Ability to vaporize or become volatile.

Volt (Symbol V): The SI unit of electrical potential

$$IV = IJC^{-1}$$

Voltameter: A vessel in which electrolysis takes place for the purpose of measuring electric current by weighing the mass of a metal.

Volume Rate: Amount of liquid formulation generally applied per unit area for effective coverage; expressed in litres per hectare or gallons per acre. In common practice, for herbicide application recommended volume rates are : high volume sprays 675-1125 litres; medium volume 225-675 litres; low volume; 50-225 litres and ultra low volume, below to litres of formulation per hectare.

Volumetric Analysis: Quan-titative analysis by measuring the volume (V_1) of a standard solution of a compound of normality (NI), that would completely react, neutralize or produce equivalence point with a known volume (V_2) of another compound of unknown normality (N_2), whose purity or quantity needs to be determined. At equivalence point $N_1V_1 = N_2V_2$; therefore N_2 and hence the purity or quantity of the unknown sample can be calculated.

Volumetric Analysis: A quantitative method involves measuring the volume of a solution of accurately known concentration that is needed to react with a solution of the substance being measured. The solution of known concentration is added in small portions from a burette. The process is called titration and the equivalence point is called the end point.

W

W: Symbol for warfarin or tungsten, a transition metal element having proton number 74 and atomic mass 183.85.

Walden inversion: Conversion of an optically active isomer into its enantiomer resulting in change of optical rotation (due to inversion of configuration) of the compound. The reaction may be a single step or multistep process.

Warfarin, $C_{19}H_{16}$O4: A colourless crystalline subs-tance, m.p. 161°C, used as a rat poison.

Water (H_2O): A colourless, odourless, tasteless liquid that freezes at 0°C and at atmospheric pressure, boils at 100°C. It is polar liquid and is the most powerful solvent.

Water of Crystallization: A particular crystalline subs-tance always has the same number of molecules of water of crystallization combined with the ions of a molecule of that substance, e.g., $CUSO_4.5H_2O$, CUl_2. $2H_2O$, $Na_2SO_4.10H_2O$, Na_2 CO_3,$10H_2O$.

Water Dispersible Product: Any formulation that is finally diluted with water, usually for spray. It includes, wettable powder (WP), emulsifiable concentrate (EC), water dispersible concentrate (WDC), water suspendible concentrate (WSC) or water dispersible slurry (WDS).

Water Gas: A mixture of carbon monoxide and hydrogen gas obtained when a blast of steam is forced up through hot bed of coke or coal. It is used as fuel and as a source of industrial preparation of hydrogen gas.

Water Get: An explosive in which dynamite is coated with a water repellant in order to slow the penetration of water.

Water Glass: Also known as sodium silicate or soluble glass, crystal like lump that range from colourless to white or greyish white and resemble glass but can be dissolved in water to form a syrupy liquid. Water glass is used as a bonding in manufacturing grinds Lands and abrasive wheels. It is a cement for glass or porcelain. It was formerly used as a for eggs.

Water of Crystallization: of water present in crystals of any compound, forming a part of the Crystal lattice as in copper sulphate pentahy-drate, $CUSO_4,5H_2O$ and sodium carbonate decahy-drate, $Na_2CO_3.10H_2O$. The water of crystallization can be removed by heating.

Water Repellant: Any of various chemicals, such as an emulsion of aluminium acetate used to make textiles, leather or other porous materials resistant to water.

Water Softener: Device for removing calcium and magnesium from water; water so treated will not form insoluble scale in pipes and tanks and will not form a precipitate with soaps or interfere with other cleaners.

Watt (Symbol W): The SI unit of powder, defined as a power of one joule per second $1W = 1JS^{-1}$.

Wave Function (Symbol y) A mathematical description of an allowed energy state, or orbital for an electron in quantum mechanical model for an atom.

Wave Length (Symbol l): The distance between the ends of one complete cycle of a wave. Wavelength is linked to the speed (C) and frequency (V) thus,

$$C = Vl$$

Wave Number: The reciprocal of the wavelength of a wave. It is the number of wave cycles in unit distance and is often used in spectroscopy. Symbol is s. The circular wave number (symbol k) is given by

$$K = 2ps$$

Wavelength (l): Distance between two successive similar points of a wave, i.e., between successive maxima (peaks) or between successive minima (troughs) of the wave, It is also defined as velocity of radiation (c) of the radiation (or wave) divided by the frequency (v) of the radiation and is generally expressed in nanometers.

Wax: A soft amorphous water repellant organic material. The term is used for fatty acid esters of monohydric alcohols.

Weber: The SI unit of magnetic flux equal to the magnetic flux that linking a circuit of one turn produces an e.m.f. of one volt when, reduced to zero at a uniform rate in one second. Symbol is Wb.

Weston Cell: A largely used standard cell Cd (Hg)/$CdSO_4$.8/3H_2O (S),Hg_2 SO_4 (S)/Hg the electrolyte is saturated solution of $CdSO_4$.

Wet Gas: Natural gas that may be gaseous or both liquid and gaseous in the reservoir and that contains condensable hydrocarbons when brought to the surface.

Wetting Agent: Chemical substance that increases the spreading and penetrating properties of a liquid by lowering its surface tension.

White Iron: Iron in which nearly all the carbon is combined with iron In the

White Spirit: A mixture of alkanes of boiling range 150°–200°C. It is used as a solvent and in the paint and varnish industry.

Wilkinsons Catalyst [Rh(Ph_3P)3Cl]: An important catalyst for homogeneous hydrogenation which also catalyses oxidation by O_2 and CO abstraction from organic derivatives.

Williamson's Continuous process: Preparation of simple ethers by dehydration of alcohols in presence of concentrated sulphuric acid at 140°C under reflux. An excess of alcohol is used and mostly dialkyl ethers of the same alkyl moiety (R) are obtained, as per the equation

$$2ROH \rightarrow ROR + H_2O$$

Williamson's Synthesis: A method used to prepare ethers by nueleophilic substitution.

$$R'Cl + RONa \rightarrow R'OR + NaCI$$

where the alkyl group R and R′ are different.

Witherite ($BaCO_3$): The white mineral form of barium carbonate. It is used as a source of Ba compounds.

Wittig Reaction: Preparation of alkenes from carbonyl compounds using ylides via a betain intermediate:

Wohler's Synthesis: Synthesis of urea from inorganic compounds. The experiment was important as it demonstrated that organic chemicalscould he made from inorganic starting materials.

X

Xanthate: Any of a class of organic salts formed by treatment of an alcohol with carbon disulphide in the presence of an alkali. The most important group of xanthates are the sodium salts produced from cellulose.

Xanthine or 2:6-dihydroxy-purine ($C_6H_4O_2N_4$): An amorphous powder obtained by reduction of uric acid with sodium amalgam. It is present in tea leaves, meat extract and urine.

Xanthommatin: Yellow biological pigment which is a metabolic product of the amino acid tryptophan.

Xanthone or Dibenzo-4 Pyrone ($Cl_3H_8O_2$): A colourless crystals, m.p. 174°C. It is prepared by the action of heat on phenyl salicylate. It is parent substance of the xanthone group of dyestuffs.

Xenon: A colourless monoatomic gas of die zero group. It exists in trace amount in air. It is used in electron tubes.

Xenon Tetrafluoride (XeF_4): A colourless crystalline solid obtained by heating a mixture of xenon and fluoride in the ratio 1:5 to 673K at 6 atmosphere pressure in a nickel container for a few hours and suddenly cooling the vapours. It dissolves in liquid HF. It dissolves in molten SF_5 or TaF_5.

Xenon Trioxide (XeO_3): A white nonvolatile compound obtained when XeF_4 or XeF_6 is hydrolysed. The solution on evaporation leaves behind a solid. It is stable in aqueous acid solution. Its aqueous solution is weakly acid and is known as xenic acid.

Xerography: A method of photographic copying in which electrostatic image is obtained on a surface coated with solenium when it is exposed to an optical image.

X-Radiations: An energetic kind of electromagnetic radiations. The range of wavelength is 10^{-11} m to 10^{-8} m.

X-Ray Crystallography: The study of the internal structure of crystals with the help of X-ray diffraction.

X-Ray Diffraction: A method employed to find out crystal structure by directing X-ray at the crystals and examining the diffraction patterns formed.

X-Ray Fluorescence: Radiation emitted by a substance when irradiated with X-rays of the appropriate wavelength.

X-Ray Fluorescence Analysis: Method of analysis in which fluorescent radiation emitted by a substance is analyzed in an X-ray spectrometer to determine the type and proportion of the elements present.

X-Ray Tube: Vacuum tube that produces X-rays by accelerating electrons to a high velocity with a high voltage field and causing them to collide with a target the anode.

Xylan: A complex polysaccharide that occurs closely associated with cellulose in plants.

Xylene or Xylol or Dimethyl Benzene, [$C_6H_4(CH_3)_2$]: A liquid resembling toluene which occurs in coal tar. It is found in three isomeric forms, a mixture of which boils at 137-140°C. It is used in the manufacture of dyes.

Xylidine or Dimethyl Aniline [$(CH_3)_2C_6H_3NH_2$]: An aromatic amine which is found in six isomeric forms, of which five are liquids, b.p. in the range of 216-230°C. It is used in the manufacture of dyes.

Y

Ylides: The of the type $Ar_3P = CRR'$, where Ar is aryl and R and R′ are alkyl groups. In these compounds carbon has carbanion character due to resonance : $Ar_3AP = CRR' = Ar_3P^+– CRR'$. These compounds can be prepared by nucleophilic attack of triphenyl-phosphene on an alkyl halide followed by abstraction of a proton by a base. Ylides are important intermediates for preparation of alkenes from ketones by wittig reaction.

Ytterbium: A soft silvery element belonging to lanthanid series. It is used to improve properties of steel.

Symbol Yb; a.n. 70; r.d. 6.97 or 6.54; m.p. 813°C; b.p. 11.94°C.

Yttrium: A silvery white metal belonging to transition series. It occurs along with lanthanides in several minerals. It is resistant to air even up to 1300 K. It is used in making alloys.

Symbol Y; a.n. 39; r.d. 4.47; m.p. 1523°C; b.p. 3337°C.

Z

Zeolite: A chemical compound with a general formula

$$Mx/n[(AlO_2)'(SiO_2)y]Z.H_2O$$

where M is metal like Na^+, K^+ or Ca^{2+}, *n* is the charge of the metal cation and Z is the number of water of hydration. It is used in softening water and for sugar refining.

Zeolite Process: A process used to soften hard water with the help of sodium zeolite

$$Na_2Al_2Si_2O_8.xH_2O$$

Zero Order: A chemical reaction in which the rate of reaction does not depend upon the concentration of a reactant, i.e., rate = $k[X]^o$.

Zero Point Energy: The energy contained by the atoms and molecules of a substance at absolute zero (OK).

Ziegler Catalysts: Catalyst capable of starting the polymerization of ethylene and propylene at normal temperature and pressure, e.g., titanium trichloride and aluminium alkyl. It is named after Carl Ziegler.

Zieloy: Zinc alloys containing approximately 1% copper and 0.1% titanium and having good ductility at normal and low temperatures.

Zinc: A transition metal occurring as the natural sulphide and carbonate.

Symbol Zn; m.p. 420°C; b.p. 908°C.

Zincate: A salt having the ion ZnO_2^{2-}

Zinc Blende (ZnS): An important ore of zinc.

Zinc Carbonate ($ZnCO_3$): A soft white powder insoluble in water. It occurs in nature as calamine or zinc spar. It is used in calamine lotion, as remedy for skin diseases and also as a white pigment.

Zinc Chloride ($ZnCl_2$): A highly deliquescent white solid with melting point 290°C. It is highly soluble in water giving acidic solution due to hydrolysis. It is used as dehydrating agent, preserving wood and paste with zinc oxide is used as dental cements.

Zinc Chromate ($ZnCrO_4$): A yellow pigment precipitate from $ZnSO_4$ solution by K_2CrO_4. It is used in rust inhibiting paints.

Zinc-Copper Couple: Metallic zinc covered with a thin film of copper by dipping zinc in copper sulphate solution. It gives hydrogen with hot water.

Zinc Fluoride ($ZnF_2.4H_2O$): Forms ZnF_2 at 100°C, formed ZnO plus aqueous HF or Zn plus HF or F_2. It is used as catalyst and wood preservative.

Zinc Cyanide [$Zn(CN)_2$]: Prepared by the interaction of zinc acetate and hydrocyanic acid. It dissolves in a solution of potassium cyanide forming a complex salt, $K_2[Zn(CN)_4]$. It is used in certain medicinal preparation.

Zinc Diethyl, $Zn(C_2H_5)_2$: An important organometallic compound of zinc with ethyl iodide. It is a volatile inflammable liquid.

Zinc Fluorosilicate ($ZnSiF_8.6H_2O$): A water soluble salt used as wood preservative and fungicide and in plastic industry.

Zinc Group: The metal zinc, cadmium and mercury constitute sub-group-II B of the periodic table. The elements of this group shows +2 oxidation state. These elements do show many properties shown by transition elements in general.

Zinc Hydroxide [Zn(OH2]: A white solid obtained by adding the requisite amount of an alkali to the solution zinc salt.

$$ZnCO_2 + 2NaOH \rightarrow Zn(OH)_2 + 2NaCl$$

It is amphoteric in nature. It is soluble in ammonia owing to the formation of the complex ion $[Zn(NH_3)_4]^{2+}$ and not ammonium zincate.

Zinc Nitrate [Zn(NO3)2]: A white crystals obtained by dissolving Zinc, its oxides, hydroxides, carbonate in nitric acid. It is deliquescent and soluble in water and alcohol.

Zinc Oxide (ZnO): A white fibrous powder obtained by heating zinc carbonate.

$$ZnCO_3 \rightarrow ZnO + CO_2\text{-}$$

It is insoluble in water on beating it charges yellow but again changes to white on cooling. It is amphoteric in nature and hence reacts with acids and alkalies. It is used as filter for rubber, in medicine, and in the manufacture of glass and pottery glazes.

Zinc Peroxide (ZnO_2): A yellowish powder obtained by treating zinc oxide with hydrogen peroxide.

$$ZnO+H_2O_2 \rightarrow ZnO_2+H_2O$$

Zinc Sulphate ($ZnSO_4$): A white crystalline solid obtained commercially by roasting the sulphide ore with air at a moderate temperature. It forms colourless rhomlic crystals having a composition $ZnSO_4.7H_2O$. It is very soluble in water, It is the most important salt of zinc. It is used in the preparation of white paint (lithophone), in zinc plating and a common starting material for other zinc compounds. A dilute solution (1/8%) is used as eye lotion.

Zinc Sulphide (ZnS): A white amorphous material is obtained from the solution of zinc salts on addition of ammonium sulphide. It is found in native as zinc blende. It is used as pigments in paints.

Zincite: Natural zinc oxide, ZnO. An important ore of zinc.

Zirconium A: lustrous metal with steel-like appearance when pure it is sufficiently soft and malleable. It is prepared industrially by the reduction of zirconium tetrachloride. It is used for making high class tough steels and the alloy ferrozirconium. ZrO_2 is used as white pigment and used in making high class refractories. Symbol Zr; a.n. 40; .d. m.p.1850°C; b.p. 4380°C.

Zirconium Dioxide or Zirconia (ZrO_2): A white crystalline insoluble compound, m.p. 2715°C. It is used as a pigment and a refractory.

Zirconium Tetrachloride (ZrO_4): A white solid obtained by heating a mixture of zirconium dioxide and carbon in chlorine. It is readily hydrolysed by water. $ZrCl_4 + 2H_2O$ ® $ZrOCl_2 + 2HCl$.

Zone Refining: The method of purifying metals by removing impurities which shows difference in solubility in the liquid and solid states of the metal. Ge, Si, and Ga are purified by this method.

Zinc Yellow: Greenish yellow pigment that is usually made by reaction of zinc oxide, potassium dichromate and sulphuric acid.

Zwitter Ion: An ion that has both a positive and negative charge on the same species. Zwitter ions occur when a molecule contains both a basic group and an acidic group as in amino acids, e.g., In glycine, $H_3N^+CH_2COO^-$. In such cases the ion formation can he regarded as an internal acid-base reaction. The zwitter ion is capable of behaving as an amphoteric at low pH (reacting with acid) it forms and at high pH, (reacting with a base) an anion (H_2NCH_2COO) is formed.